PERGAMON INTERNATIONAL LIBRARY
of Science, Technology, Engineering and Social Studies
*The 1000-volume original paperback library in aid of education,
industrial training and the enjoyment of leisure*
Publisher: Robert Maxwell, M.C.

Nuclear Energy

SECOND EDITION

in SI/metric Units

Pergamon Unified
Engineering Series

Volume 22

THE PERGAMON TEXTBOOK
INSPECTION COPY SERVICE

An inspection copy of any book published in the Pergamon International Library will
gladly be sent to academic staff without obligation for their consideration for course adoption
or recommendation. Copies may be retained for a period of 60 days from receipt and returned
if not suitable. When a particular title is adopted or recommended for adoption for class use
and the recommendation results in a sale of 12 or more copies, the inspection copy may be
retained with our compliments. The Publishers will be pleased to receive suggestions for
revised editions and new titles to be published in this important International Library.

Pergamon
Unified Engineering
Series

GENERAL EDITORS

Thomas F. Irvine, Jr.
State University of New York at Stony Brook

James P. Hartnett
University of Illinois at Chicago Circle

EDITORS

William F. Hughes
Carnegie-Mellon University

Arthur T. Murphy
Widener College

Daniel Rosenthal
University of California, Los Angeles

SECTIONS

Continuous Media Section
Engineering Design Section
Engineering Systems Section
Humanities and Social Sciences Section
Information Dynamics Section
Materials Engineering Section
Engineering Laboratory Section

Nuclear Energy

An Introduction to the Concepts, Systems, and Applications of Nuclear Processes

SECOND EDITION (in SI/metric Units)

Raymond L. Murray
Nuclear Engineering Department,
North Carolina State University,
Raleigh, North Carolina, 27650
USA

PERGAMON PRESS

Oxford · New York · Toronto · Sydney · Paris · Frankfurt

U.K.	Pergamon Press Ltd., Headington Hill Hall, Oxford OX3 0BW, England
U.S.A.	Pergamon Press Inc., Maxwell House, Fairview Park, Elmsford, New York 10523, U.S.A.
CANADA	Pergamon of Canada, Suite 104, 150 Consumers Road, Willowdale, Ontario M2J 1P9, Canada
AUSTRALIA	Pergamon Press (Aust.) Pty. Ltd., P.O. Box 544, Potts Point, N.S.W. 2011, Australia
FRANCE	Pergamon Press SARL, 24 rue des Ecoles, 75240 Paris, Cedex 05, France
FEDERAL REPUBLIC OF GERMANY	Pergamon Press GmbH, 6242 Kronberg-Taunus, Hammerweg 6, Federal Republic of Germany

First edition 1975

Second edition 1980

British Library Cataloguing in Publication Data

Murray, Raymond LeRoy
Nuclear energy. — 2nd ed. — (Pergamon unified engineering series; vol.22).
1. Atomic power
I. Title
621.48   TK9145   79—41701

ISBN 0-08-024751-2 hardcover
ISBN 0-08-024750-4 flexicover

Printed and bound in Great Britain by
William Clowes (Beccles) Limited, Beccles and London

To Elizabeth

Preface to the Second Edition

In the period since *Nuclear Energy* was written, there have been several significant developments. The Arab oil embargo with its impact on the availability of gasoline alerted the world to the increasing energy problem. The nuclear industry has experienced a variety of problems including difficulty in financing nuclear plants, inflation, inefficiency in construction, and opposition by various intervening organizations. The accident at Three Mile Island raised concerns in the minds of the public and led to a new scrutiny of safety by government and industry.

Two changes in U.S. national administration of nuclear energy have occurred: (a) the reassignment of responsibilities of the Atomic Energy Commission to the Nuclear Regulatory Commission (NRC) and the Energy Research and Development Administration (ERDA) which had a charge to develop all forms of energy, not just nuclear, (b) the absorption of ERDA and the Federal Energy Agency into a new Department of Energy. Recently, more attention has been paid to the problem of proliferation of nuclear weapons, with new views on fuel reprocessing, recycling, and the use of the breeder reactor. At the same time, several nuclear topics have become passé.

The rapidly changing scene thus requires that we update *Nuclear Energy*, without changing the original intent as described in the earlier Preface. In preparing the new version, we note in the text and in the Appendix the transition in the U.S. to SI units. New values of data on materials are included, e.g. atomic masses, cross sections, half lives, and radiations. Some new problems have been added. The Appendix has been expanded to contain useful constants and the answers to most of the problems. Faculty users are encouraged to secure a copy of the *Solution Manual* from the publisher.

Thanks are due Dr. Ephraim Stam for his careful scrutiny of the draft and for his fine suggestions. Thanks also go to Mary C. Joseph and Rashid Sultan for capable help with the manuscript.

Raleigh, North Carolina   RAYMOND L. MURRAY

Preface to the First Edition

The future of mankind is inextricable from nuclear energy. As the world population increases and eventually stabilizes, the demands for energy to assure adequate living conditions will severely tax available resources, especially those of fossil fuels. New and different sources of energy and methods of conversion will have to be explored and brought into practical use. The wise use of nuclear energy, based on understanding of both hazards and benefits, will be required to meet this challenge to existence.

This book is intended to provide a factual description of basic nuclear phenomena, to describe devices and processes that involve nuclear reactions, and to call attention to the problems and opportunities that are inherent in a nuclear age. It is designed for use by anyone who wishes to know about the role of nuclear energy in our society or to learn nuclear concepts for use in professional work.

In spite of the technical complexity of nuclear systems, students who have taken a one-semester course based on the book have shown a surprising level of interest, appreciation, and understanding. This response resulted in part from the selectivity of subject matter and from efforts to connect basic ideas with the "real world," a goal that all modern education must seek if we hope to solve the problems facing civilization.

The sequence of presentation proceeds from fundamental facts and principles through a variety of nuclear devices to the relation between nuclear energy and peaceful applications. Emphasis is first placed on energy, atoms and nuclei, and nuclear reactions, with little background required. The book then describes the operating principles of radiation equipment, nuclear reactors, and other systems involving nuclear processes, giving quantitative information wherever possible. Finally, attention is directed to the subjects of radiation protection, beneficial usage of radiation, and the connection between energy resources and human progress.

The author is grateful to Dr. Ephraim Stam for his many suggestions on technical content, to Drs. Claude G. Poncelet and Albert J. Impink, Jr. for their careful review, to Christine Baermann for her recommendations on style and clarity, and to Carol Carroll for her assistance in preparation of the manuscript.

Raleigh, North Carolina   RAYMOND L. MURRAY

Contents

The Author

Raymond L. Murray (Ph.D. University of Tennessee) is professor in the Department of Nuclear Engineering at North Carolina State University at Raleigh. His professional interests are in nuclear reactor design analysis, reactor and radiation safety, and nuclear engineering education.

Dr. Murray is the author of several textbooks in physics and nuclear technology and has written many research papers in reactor analysis. He serves as a consultant to the nuclear industry. Dr. Murray's career in the nuclear field began in 1942 with the Manhattan Project at Berkeley and continued at Oak Ridge. In 1950 he helped found the first university nuclear engineering programs. He is Fellow of the American Physical Society and of the American Nuclear Society and is a member of several other scientific and engineering societies.

Part I   Basic Concepts

In the study of the practical applications of nuclear energy, we must take account of the properties of individual particles of matter—their "microscopic" features—as well as the character of matter in its ordinary form, a "macroscopic" (large-scale) view. Examples of the small-scale properties are masses of atoms or nuclear particles, their effective sizes for interaction with each other, or the number of particles in a certain volume. The combined behavior of large numbers of individual particles is expressed in terms of properties such as mass density, charge density, electrical conductivity, thermal conductivity, and elastic constants. We continually seek consistency between the microscopic and macroscopic views.

Since all processes involve interactions of particles, it is necessary that we develop a background of understanding of the basic physical facts and principles that govern such interactions. In Part I, we shall examine the concept of energy, describe the models of atomic and nuclear structure, discuss radioactity and nuclear reactions in general, review the ways radiation reacts with matter, and concentrate on two important nuclear processes—fission and fusion.

1

Energy and States of Matter

Our material world is composed of many substances distinguished by their chemical, mechanical, and electrical properties. They are found in nature in various physical states—the familiar solid, liquid, and gas, along with the ionic "plasma." However, the apparent diversity of kinds and forms of material is reduced by the knowledge that there are only a little over 100 distinct chemical elements and that the chemical and physical features of substances depend merely on the strength of force bonds between atoms.

In turn, the distinctions between the elements of nature arise from the number and arrangement of basic particles—electrons, protons, and neutrons. At both the atomic and nuclear levels, the structure of elements is determined by internal forces and energy.

1.1 FORCES AND ENERGY

There is a limited number of basic forces—gravitational, electrostatic, electromagnetic, and nuclear. Associated with each of these is the ability to do work. Thus energy in different forms may be stored, released, transformed, transferred, and "used" in both natural processes and man-made devices. It is often convenient to view nature in terms of only two basic entities—particles and energy. Even this distinction can be removed, since we know that matter can be converted into energy and vice versa.

Let us review some principles of physics needed for the study of the release of nuclear energy and its conversion into thermal and electrical

3

form. We recall that if a constant force F is applied to an object to move it a distance s, the amount of work done is the product Fs. As a simple example, we pick up a book from the floor and place it on a table. Our muscles provide the means to lift against the force of gravity on the book. We have done work on the object, which now possesses stored energy (potential energy), because it could do work if allowed to fall back to the original level. Now a force F acting on a mass m provides an acceleration a, given by Newton's law $F = ma$. Starting from rest, the object gains a speed v, and at any instant has energy of motion (kinetic energy) in amount $E_k = \frac{1}{2} mv^2$. For objects falling under the force of gravity, we find that the potential energy is reduced as the kinetic energy increases, but the sum of the two types remains constant. This is an example of the principle of conservation of energy. Let us apply this principle to a practical situation and perform some illustrative calculations.

As we know, falling water provides one primary source for generating electrical energy. In a hydroelectric plant, river water is collected by a dam and allowed to fall through a considerable height. The potential energy of water is thus converted into kinetic energy. The water is directed to strike the blades of a turbine, which turns an electric generator. The force of gravity on a mass of water is $F = mg$, where g is the acceleration of gravity, 9.8 m/sec². At the top of the dam of height h, the potential energy is $E_p = Fh$ or mgh. For instance, each kilogram at level 50 meters has energy (1)(9.8)(50) = 490 joules (J). Ignoring friction effects, this amount of energy in kinetic form would appear at the bottom. The speed of the water would be $v = \sqrt{2E_k/m} = 31.3$ m/sec.

Energy takes on various forms, classified according to the type of force that is acting. The water in the hydroelectric plant experiences the force of gravity, and thus gravitational energy is involved. It is transformed into mechanical energy of rotation in the turbine, which then is converted to electrical energy by the generator. At the terminals of the generator, there is an electrical potential difference, which provides the force to move charged particles (electrons) through the network of the electrical supply system. The electrical energy may then be converted into mechanical energy as in motors, or into light energy as in lightbulbs, or into thermal energy as in electrically heated homes, or into chemical energy as in a storage battery.

The automobile also provides familiar examples of energy transformations. The burning of gasoline releases the chemical energy of the fuel in the form of heat, part of which is converted to energy of motion of mechanical parts, while the rest is transferred to the atmosphere and highway. Electricity is provided by the car's generator for control and lighting. In each of these examples, energy is changed from one form to another, but is not destroyed. The conversion of heat to other forms of energy is governed by two laws, the first and second laws of thermodynamics. The first states that energy is conserved; the second specifies inherent limits on the efficiency of energy conversion.

Energy can be classified according to the primary source. We have already noted two sources of energy: falling water and the burning of the chemical fuel gasoline, which is derived from petroleum, one of the main fossil fuels. To these we can add solar energy, the energy from winds, tides, or other sea motion, and heat from within the earth. Finally, we have energy from nuclear reactions, i.e., the "burning" of nuclear fuel.

1.2 THERMAL ENERGY

Of special importance to us is thermal energy, as the form most readily available from the sun, from burning of ordinary fuels, and from the fission process. First we recall that a simple definition of the temperature of a substance is the number read from a measuring device such as a thermometer in intimate contact with the material. If energy is supplied, the temperature rises, e.g., energy from the sun warms the air during the day. Each material responds to the supply of energy according to its internal molecular or atomic structure, characterized on a macroscopic scale by the specific heat c. If an amount of thermal energy added to one gram of the material is Q, the temperature rise, ΔT, is Q/c. The value of the specific heat for water is $c = 4.18$ J/g-°C and thus it requires 4.18 joules of energy to raise the temperature of one gram of water by one degree Celsius (1°C).

From our modern knowledge of the atomic nature of matter, we readily appreciate the idea that energy supplied to a material increases the motion of the individual particles of the substance. Temperature can thus be related to the average kinetic energy of the atoms. For example, in a gas such as air, the average energy of translational motion of the molecules $\bar{E}$ is directly proportional to the temperature T, through the relation $\bar{E} = \frac{3}{2} kT$, where k is Boltzmann's constant, 1.38×10^{-23} J/°K. (Note that the Kelvin scale has the same spacing of degrees as does the Celsius scale, but its zero is at -273°C.)

To gain an appreciation of molecules in motion, let us find the typical speed of oxygen molecules at room temperature 20°C or 293°K. The molecular weight is 32, and since one unit of atomic weight corresponds to 1.66×10^{-27} kg, the mass of the oxygen (O_2) molecule is 5.3×10^{-26} kg. Now

$$\bar{E} = \tfrac{3}{2}(1.38 \times 10^{-23})(293) = 6.1 \times 10^{-21} \text{ J},$$

and thus the speed is

$$v = \sqrt{2\bar{E}/m} = \sqrt{2(6.1 \times 10^{-21})/(5.3 \times 10^{-26})} \cong 479 \text{ m/sec}.$$

Closely related to energy is the physical entity *power*, which is the rate at which work is done. To illustrate, suppose that the flow of water in a hydroelectric plant were 2×10^6 kg/sec. The corresponding energy per second is $(2 \times 10^6)(490) = 9.8 \times 10^8$ J/sec. For convenience, the unit joules per second is called the watt (W). Our plant thus involves 9.8×10^8 W. We can conveniently express this in kilowatts ($1 \text{ kW} = 10^3 \text{ W}$) or megawatts ($1 \text{ MW} = 10^6 \text{ W}$).

For most purposes we shall employ the metric system of units, more precisely designated as SI, Système Internationale. In this system the base units, as described in the Federal Register, December 10, 1976, are the kilogram (kg) for mass, the meter (m) for length, the second (s) for time, the mole (mol) for amount of substance, the ampere (A) for electric current, the kelvin (K) for thermodynamic temperature and the candela (cd) for luminous intensity. However, for understanding of the earlier literature, one requires a knowledge of other systems. The Appendix includes a table of useful conversions from British to SI units.

In dealing with forces and energy at the level of molecules, atoms, and nuclei, it is conventional to use another energy unit, the *electron-volt* (eV). Its origin is electrical in character, being the amount of energy that would be imparted to an electron (charge 1.60×10^{-19} coulombs) if it were accelerated through a potential difference of 1 volt. Since the work done on 1 coulomb would be 1 J, we see that $1 \text{ eV} = 1.60 \times 10^{-19}$ J. The unit is of convenient size for describing atomic reactions. For instance, to remove the one electron from the hydrogen atom requires 13.5 eV of energy. However, when dealing with nuclear forces, which are very much larger than atomic forces, it is preferable to use the million-electron-volt unit (MeV). To separate the neutron from the proton in the nucleus of heavy hydrogen, for example, requires an energy of about 2.2 MeV, i.e., 2.2×10^6 eV.

1.3 RADIANT ENERGY

Another form of energy is electromagnetic or radiant energy. We recall that this energy may be released by heating of solids, as in the wire of a lightbulb, or by electrical oscillations, as in radio or television transmitters, or by atomic interactions, as in the sun. The radiation can be viewed in either of two ways—as a combination of electric and magnetic waves traveling through space, or as a compact moving uncharged particle—the photon, which is a bundle of pure energy, effectively having mass only by virtue of its motion. The choice of model—wave or particle—is dependent on the process under study. Regardless of its origin, all radiation can be characterized by its frequency, which is related to speed and wavelength. Letting c be the speed of light, λ its wavelength and ν its frequency, we have $c = \lambda\nu$.† For example, if c in a vacuum is 3×10^8 m/sec, yellow light of wavelength 5.89×10^{-7} m has a frequency of 5.1×10^{14} sec^{-1}. X-rays and gamma rays are electromagnetic radiation arising from the interactions of atomic and nuclear particles, respectively, and are different from light from other sources only in the range of energy represented.

In order to appreciate the relation of states of matter, atomic and nuclear interactions, and energy, let us visualize an experiment in which we supply energy to a sample of water from a source of energy that is as large and as sophisticated as we wish. Thus we increase the degree of internal motion and eventually dissociate the material into its most elementary components. Suppose, Fig. 1.1, that the water is initially as ice at nearly absolute zero temperature, where water (H_2O) molecules are essentially at rest. As we add thermal energy to increase the temperature to 0°C or 32°F, molecular movement increases to the point where the ice melts to become liquid water, which can flow rather freely. To cause a change from the solid state to the liquid state, a definite amount of energy (the heat of fusion) is required. In the case of water, this latent heat is 334 J/g. In the temperature range in which water is liquid, thermal agitation of the molecules permits some evaporation from the surface. At the boiling point, 100°C or 212°F at atmospheric pressure, the liquid turns into the gaseous form as steam. Again, energy is required to cause the change of state, with a heat of vaporization of 2258 J/g. Further heating, using special high temperature equipment, causes dissociation

---

†We shall have need of both Roman and Greek characters, identifying the latter by name the first time they are used, thus λ (lambda) and ν (nu). The reader must be wary of symbols used for more than one quantity.

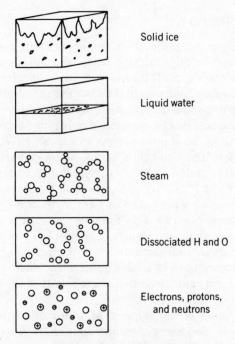

Solid ice

Liquid water

Steam

Dissociated H and O

Electrons, protons, and neutrons

Fig. 1.1.   Effect of added energy.

of water into atoms of hydrogen (H) and oxygen (O). By electrical means, electrons can be removed from hydrogen and oxygen atoms, leaving a mixture of charged ions and electrons. Through nuclear bombardment, the oxygen nucleus can be broken into smaller nuclei, and in the limit of temperatures in the billions of degrees, the material can be decomposed into an assembly of electrons, protons, and neutrons.

1.4 THE EQUIVALENCE OF MATTER AND ENERGY

The connection between energy and matter is provided by Einstein's theory of special relativity. It predicts that the mass of any object increases with its speed. Letting the mass when the object is at rest be m_0, the "rest mass," and letting m be the mass when it is at speed v, and noting that the speed of light in a vacuum is $c = 3 \times 10^8$ m/sec, then

$$m = \frac{m_0}{\sqrt{1 - (v/c)^2}}.$$

For motion at low speed (e.g., 500 m/sec), the mass is almost identical to the rest mass, since v/c and its square are very small. Although the theory has the status of natural law, its rigor is not required except for particle motion at high speed, i.e., when v is at least a few percent of c. The relation shows that a material object can have a speed no higher than c.

The kinetic energy imparted to a particle by the application of force according to Einstein is

$$E_k = (m - m_0)c^2.$$

(For low speeds, $v \ll c$, this is approximately $\frac{1}{2}m_0v^2$, the classical relation.)

The implication of Einstein's formula is that any object has an energy $E_0 = m_0c^2$ when at rest (its "rest energy"), and a total energy $E = mc^2$, the difference being E_k the kinetic energy. Let us compute the rest energy for an electron of mass 9.1×10^{-31} kg.

$$E_0 = m_0c^2 = (9.1 \times 10^{-31})(3.0 \times 10^8)^2 = 8.2 \times 10^{-14} \text{ J}$$

or

$$E_0 = \frac{8.2 \times 10^{-14} \text{ J}}{1.60 \times 10^{-13} \text{ J/MeV}} = 0.51 \text{ MeV}.$$

For one unit of atomic mass, 1.66×10^{-27} kg, which is close to the mass of a hydrogen atom, the corresponding energy is 931 MeV.

Thus we see that matter and energy are equivalent, with the factor c^2 relating the amounts of each. This suggests that matter can be converted into energy and that energy can be converted into matter. Although Einstein's relation is completely general, it is especially important in calculating the release of energy by nuclear means. We find that *the energy yield from a kilogram of nuclear fuel is more than a million times that from chemical fuel.* To prove this startling statement, we first find the result of complete transformation of a kilogram of matter into energy, viz. $(1 \text{ kg})(3.0 \times 10^8 \text{ m/sec})^2 = 9 \times 10^{16} \text{ J}$. The nuclear fission process, as one method of converting mass into energy, is relatively inefficient, since the "burning" of 1 kg of uranium involves the conversion of only 0.87 g of matter into energy. This corresponds to about 7.8×10^{13} J/kg of the uranium consumed. The enormous magnitude of this energy release can be appreciated only by comparison with the energy of combustion of a familiar fuel such as gasoline, 5×10^7 J/kg. The ratio of these numbers, 1.5×10^6, reveals the tremendous difference between nuclear and chemical energies.

1.5 ENERGY AND THE WORLD

All of the activities of human beings depend on energy, as we realize when we consider the dimensions of the world's energy problem. The efficient production of food requires machines, fertilizer, and water, each using energy in different ways. Energy is vital to transportation, protection against the weather, and the manufacturing of all goods. An adequate long-term supply of energy is therefore essential for man's survival. The energy problem or energy crisis has many dimensions — the increasing cost to acquire fuels as they become more scarce; the effects on safety and health of the byproducts of energy consumption; the inequitable distribution of energy resources among regions and nations; the discrepancies between current energy usage and human expectations throughout the world.

1.6 SUMMARY

Associated with each basic type of force is an energy, which may be transformed to another form for practical use. The addition of thermal energy to a substance causes an increase in temperature and the amount of particle motion. Electromagnetic radiation arising from electrical devices, atoms, or nuclei may be considered as composed of waves or of photons. Matter can be converted into energy and vice versa, according to Einstein's formula $E = mc^2$. The energy of nuclear fission is millions of times as large as that from chemical reactions. Energy is fundamental to all of man's endeavors and indeed to his survival.

1.7 PROBLEMS

1.1. Find the kinetic energy of a basketball player of mass 75 kg as he moves down the floor at a speed of 8 m/sec.

1.2. Recalling the conversion formulas for temperature,

$$C = \frac{5}{9}(F - 32)$$

and

$$F = \frac{9}{5}C + 32$$

where C and F are degrees in respective systems, convert each of the following: 68°F, 500°F, –273°C, 1000°C.

1.3. If the specific heat of iron is 0.45 J/g-°C how much energy is required to bring 0.5 kg of iron from 0°C to 100°C?

1.4. Find the speed corresponding to the average energy of nitrogen gas molecules (N_2, 28 units of atomic weight) at room temperature.

1.5. Find the power in kilowatts of an auto rated at 200 horsepower. In a drive for 4 hr at average speed 45 mph, how many kWhr of energy are required?

1.6. Find the frequency of a gamma ray photon of wavelength 1.5×10^{-12} m.

1.7. Verify that the mass of a typical slowly moving object is not much greater than its mass at rest (e.g., a car with $m_0 = 1000$ kg moving at 20 m/sec) by finding the number of *grams* of mass increase.

1.8. Noting that the electron-volt is 1.60×10^{-19} J, how many joules are released in the fission of one uranium nucleus, which yields 190 MeV?

1.9. Applying Einstein's formula for the equivalence of mass and energy, $E = mc^2$, where $c = 3 \times 10^8$ m/sec, the speed of light, how many kilograms of matter are converted into energy in Problem 1.8?

1.10. If the atom of uranium-235 has a mass of $(235)(1.66 \times 10^{-27})$ kg, what amount of equivalent energy does it have?

1.11. Using the results of Problems 1.8, 1.9, and 1.10, what fraction of the mass of a U-235 nucleus is converted into energy when fission takes place?

1.12. Show that to obtain a power of 1 W from fission of uranium, it is necessary to cause 3.3×10^{10} fission events per second.

1.13. (a) If the fractional mass increase due to relativity is $\Delta E/E_0$, show that

$$v/c = \sqrt{1 - (1 + \Delta E/E_0)^{-2}}$$

(b) At what fraction of the speed of light does a particle have a mass that is 1% higher than the rest mass? 10%? 100%?

1.14. The heat of combustion of hydrogen by the reaction $2H + O = H_2O$ is quoted to be 34.18 kilogram calories per gram of hydrogen. (a) Find how many Btu per pound this is using the conversions 1 Btu = 0.252 kcal, 1 lb = 454 grams. (b) Find how many joules per gram this is noting 1 cal = 4.185 J. (c) Calculate the heat of combustion in eV per H_2 molecule.

2

Atoms and Nuclei

A complete understanding of the microscopic structure of matter and the exact nature of the forces acting is yet to be realized. However, excellent models have been developed to predict behavior to an adequate degree of accuracy for most practical purposes. These models are descriptive or mathematical, often based on analogy with large-scale processes, on experimental data, or on advanced theory.

2.1 ATOMIC THEORY

The most elementary concept is that matter is composed of individual particles—atoms—that retain their identity as elements in ordinary physical and chemical interactions. Thus a collection of helium atoms that forms a gas has a total weight that is the sum of the weights of the individual atoms. Also, when two elements combine to form a compound (e.g., if carbon atoms combine with oxygen atoms to form carbon monoxide molecules), the total weight of the new substance is the sum of the weights of the original elements.

There are more than 100 known elements. Most are found in nature; some are artificially produced. Each is given an atomic number in the periodic table of the elements—examples are hydrogen (H) 1, helium (He) 2, oxygen (O) 8, and uranium (U) 92. The symbol Z is given to the atomic number, which is also the number of electrons in the atom and determines its chemical properties.

Generally, the higher an element is in the periodic table the greater is its weight, either as an individual particle or as an assembly of particles on the standard scale. The atomic weights, labeled M, of the example elements above are approximately H 1.008, He 4.003, O 16.00, and U 238.0. These values represent the number of grams of the element in a sample that contains a specific number of particles—Avogadro's number, N_a, 6.02×10^{23}. We can easily find the number of atoms per cubic centimeter in a substance if its density ρ in grams per cubic centimeter is known. For example, if we had a container of helium gas with density $0.00018 \, g/cm^3$, each cubic centimeter would contain a fraction 0.00018/4.003 of Avogadro's number of helium atoms, i.e., 2.7×10^{19}. This procedure can be expressed as a convenient formula for finding N, the number per cubic centimeter for any material:

$$N = \frac{\rho}{M} N_a.$$

Thus in uranium with density $19 \, g/cm^3$, we find $N = (19/238)(6.02 \times 10^{23}) = 0.048 \times 10^{24} \, cm^{-3}$. The relation holds for compounds as well, if M is taken as the molecular weight. In water, H_2O, with $\rho = 1.0 \, g/cm^3$ and $M = 2(1.008) + 16.00 \cong 18.0$, we have $N = (1/18)(6.02 \times 10^{23}) = 0.033 \times 10^{24} \, cm^{-3}$. (The use of numbers times 10^{24} will turn out to be convenient later.)

2.2 GASES

Substances in the gaseous state are described approximately by the perfect gas law, relating pressure, volume, and absolute temperature,

$$pV = nkT,$$

where n is the number of particles and k is Boltzmann's constant. An increase in the temperature of the gas due to heating causes greater molecular motion, which results in an increase of particle bombardment of a container wall and thus of pressure on the wall. The particles of gas, each of mass m, have a variety of speeds v in accord with Maxwell's gas theory, as shown in Fig. 2.1. The most probable speed, at the peak of this maxwellian distribution, is dependent on temperature according to the relation

$$v_p = \sqrt{\frac{2kT}{m}}.$$

The kinetic theory of gases provides a basis for calculating properties such as the specific heat. Using the fact from Chapter 1 that the average

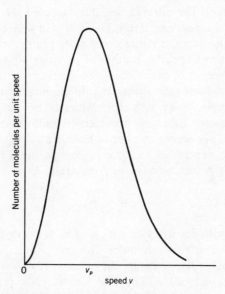

Fig. 2.1.   Distribution of molecular speeds.

energy of gas molecules is proportional to the temperature, $\bar{E} = \frac{3}{2}kT$, we can deduce that the specific heat of a gas consisting only of atoms is $c = \frac{3}{2}k/m$, where m is the mass of one atom. We thus see an intimate relationship between mechanical and thermal properties of materials.

2.3   THE ATOM AND LIGHT

It is well known that the color of a heated solid or gas changes as the temperature is increased, tending to go from the red end of the visible region toward the blue end, i.e., from long wavelengths to short wavelengths. The measured distribution of light among the different wavelengths at a certain temperature can be explained by the assumption that light is in the form of photons. These are absorbed and emitted with definite amounts of energy E that are proportional to the frequency ν, according to

$$E = h\nu,$$

where h is Planck's constant 6.63×10^{-34} J-sec. For example, the energy corresponding to a frequency of 5×10^{14} is $(6.63 \times 10^{-34})(5 \times 10^{14}) = 3.3 \times 10^{-19}$ J, which is seen to be a very minute amount of energy.

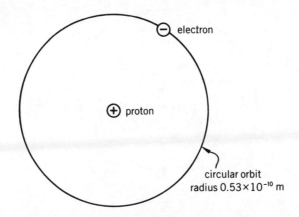

Fig. 2.2. Hydrogen atom.

The emission and absorption of light from incandescent hydrogen gas was first explained by Bohr, using a novel model of the hydrogen atom. He assumed that the atom consisted of a single electron moving at constant speed in a circular orbit about a nucleus—the proton—as sketched in Fig. 2.2. Each particle has an electric charge of 1.6×10^{-19} coulombs, but the proton has a mass that is 1836 times that of the electron. The radius of the orbit is set by the equality of electrostatic force, attracting the two charges toward each other, to centripetal force, required to keep the electron on a circular path. If energy is supplied to the hydrogen atom from the outside, the electron is caused to jump to a larger orbit of definite radius. At some later time, the electron falls back spontaneously to the original orbit, and energy is released in the form of a photon of light. The energy of the photon $h\nu$ is equal to the difference between energies in the two orbits. The smallest orbit has a radius $R_1 = 0.53 \times 10^{-10}$ m, while the others have radii increasing as the square of integers (called quantum numbers). Thus if n is $1, 2, 3, \ldots$, the radius of the nth orbit is $R_n = n^2 R_1$. Figure 2.3 shows the allowed electron orbits in hydrogen. The energy of the atom system when the electron is in the first orbit is $E_1 = -13.5$ eV, where the negative sign means that energy must be supplied to remove the electron to a great distance and leave the hydrogen as a positive ion. The energy when the electron is in the nth orbit is $E_n = E_1/n^2$. The various discrete levels are sketched in Fig. 2.4.

The electronic structure of the other elements is described by the shell model, in which a limited number of electrons can occupy a given orbit or

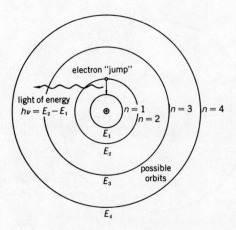

Fig. 2.3. Electron orbits in hydrogen (Bohr theory).

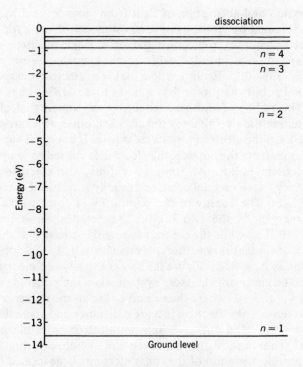

Fig. 2.4. Energy levels in hydrogen atom.

shell. The atomic number Z is unique for each chemical element, and represents both the number of positive charges on the central massive nucleus of the atom and the number of electrons in orbits around the nucleus. The maximum allowed numbers of electrons in orbits as Z increases for the first few shells are 2, 8, and 18. The chemical behavior of elements is determined by the number of electrons in the outermost or valence shell. For example, oxygen with $Z = 8$ has two electrons in the inner shell, six in the outer. Thus oxygen has an affinity for elements with two electrons in the valence shell. The formation of molecules from atoms by electron sharing is illustrated by Fig. 2.5, which shows the water molecule.

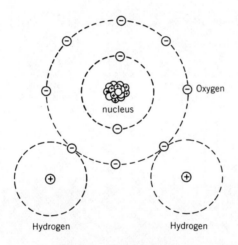

Fig. 2.5.   Water molecule.

Ordinary light as in the visible range is a mixture of many frequencies, phases, and directions since chaotic atomic processes are involved. In contrast, light from a laser (*l*ight *a*mplification by *s*timulated *e*mission of *r*adiation) consists of a very intense direct beam of one color. Energy is supplied to a lasing material such as carbon dioxide or glass containing neodymium to excite the electrons in the atoms to higher energy states. When a photon of light of the right frequency is introduced, an excited atom emits a second photon. The two photons, in phase and moving in the same direction, create four photons, and so on, until an avalanche of light is produced. Beams from lasers have many applications in science and industry. Nuclear uses are discussed in Sections 10.5 and 16.4.

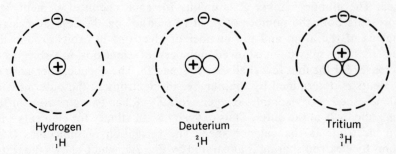

Fig. 2.6.   Isotopes of hydrogen.

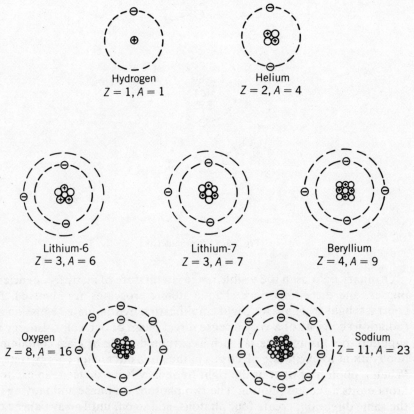

Fig. 2.7.   Atomic and nuclear structure.

2.4 NUCLEAR STRUCTURE

Most elements are composed of particles of different weight, called isotopes. For instance, hydrogen has three isotopes of weights in proportion 1, 2, and 3—ordinary hydrogen, heavy hydrogen (deuterium), and tritium. Each has atomic number $Z = 1$ and the same chemical properties, but they differ in the composition of the central nucleus, where most of the weight resides. The nucleus of ordinary hydrogen is the positively charged proton; the deuteron consists of a proton plus a neutron, a neutral particle of weight very close to that of the proton; the triton contains a proton plus two neutrons. To distinguish isotopes, we identify the mass number A, as the total number of nucleons, the heavy particles in the nucleus. A complete shorthand description is given by the chemical symbol with superscript A value and subscript Z value, e.g., ^1_1H, ^2_1H, ^3_1H. Figure 2.6 shows the nuclear and atomic structure of the three hydrogen isotopes. Each has one electron in the outer shell, in accord with the Bohr theory described earlier.

The structure of some of the lighter elements and isotopes is sketched in Fig. 2.7. In each case, the atom is neutral because the negative charge of the Z electrons in the outer shell balances the positive charge of the Z protons in the nucleus. The symbols for these are ^1_1H, ^4_2He, ^6_3Li, ^7_3Li, ^9_4Be, $^{16}_8O$, and $^{23}_{11}Na$. In addition to the atomic number Z and the mass number A, we often need to write the neutron number N, which is, of course, $A - Z$. For the set of isotopes listed, N is 0, 2, 3, 4, 5, 8, and 12, respectively.

When we study nuclear reactions, it is convenient to let the neutron be represented by the symbol ^1_0n, implying a mass comparable to that of hydrogen ^1_1H, but with no electronic charge, $Z = 0$. Similarly, the electron is represented by $^0_{-1}e$, suggesting nearly zero mass in comparison with that of hydrogen, but with negative charge. An identification of isotopes frequently used in qualitative discussion consists of the element name and its A value, thus sodium-23 and uranium-235, or even more simply Na-23 and U-235.

2.5 SIZES AND MASSES OF NUCLEI

The dimensions of nuclei are found to be very much smaller than those of atoms. Whereas the hydrogen atom has a radius of about 5×10^{-9} cm, its nucleus has a radius of only about 10^{-13} cm. Since the proton weight is much larger than the electron weight, the nucleus is extremely dense. The nuclei of other isotopes may be viewed as closely packed particles of matter—neutrons and protons—forming a sphere whose volume, $\frac{4}{3}\pi R^3$, depends on A, the number of nucleons. A useful rule of thumb to calculate radii of nuclei is

$$R(\text{cm}) = 1.4 \times 10^{-13} A^{1/3}.$$

Since A ranges from 1 to about 250, we see that all nuclei are smaller than 10^{-12} cm.

The masses of atoms labeled M are compared on a scale in which an isotope of carbon $^{12}_{6}C$ has a mass of exactly 12. For $^{1}_{1}H$, the atomic mass is $M = 1.007825$, for $^{2}_{1}H$, $M = 2.014102$, and so on. The atomic mass of the proton is 1.007277, of the neutron 1.008665, the difference being only about 0.1%. The mass of the electron on this scale is 0.000549. A list of atomic masses appears in the Appendix.

The atomic mass unit (amu), as $\frac{1}{12}$ the mass of $^{12}_{6}C$, corresponds to an actual mass of 1.66×10^{-24} g. To verify this, merely divide 1 g by Avogadro's number 6.02×10^{23}. It is easy to show that 1 amu is also equivalent to 931 MeV. We can calculate the actual masses of atoms and nuclei by multiplying the mass in atomic mass units by the mass of 1 amu. Thus the mass of the neutron is $(1.008665)(1.66 \times 10^{-24}) = 1.67 \times 10^{-24}$ g.

2.6 BINDING ENERGY

The force of electrostatic repulsion between like charges, which varies inversely as the square of their separation, would be expected to be so large that nuclei could not be formed. The fact that they do exist is evidence that there is an even larger force of attraction. This nuclear force acts only when the nucleons are very close to each other and binds them into a compact stable structure. Associated with the net force is a potential energy of binding. To disrupt a nucleus and separate it into its component nucleons, energy must be supplied from the outside. Recalling Einstein's relation between mass and energy, this is the same as saying that mass must be supplied to the nucleus. A given nucleus is lighter than the sum of its separate nucleons, the difference being the binding mass-energy. Let the mass of an atom including nucleus and external electrons be M, and let m_n and m_H be the masses of the neutron and the proton plus matching electron. Then the binding energy is

$$B = \text{total mass of separate particles} - \text{mass of the atom}$$

or

$$B = Nm_n + Zm_H - M.$$

(Neglected in this relation is a small energy of atomic or chemical binding.) Let us calculate B for tritium, the heaviest hydrogen atom. Figure 2.8 shows the dissociation that would take place if a sufficient energy were provided. Now $Z = 1$, $N = 2$, $m_n = 1.008665$, $m_H = 1.007825$, and $M = 3.016049$. Then

$$B = 2(1.008665) + 1(1.007825) - 3.016049$$
$$B = 0.009106 \text{ amu.}$$

Converting by use of the relation 1 amu = 931 MeV, the binding energy is $B = 8.48$ MeV. Calculations such as these are required for several purposes—to compare the stability of one nucleus with that of another, to find the energy release in a nuclear reaction, and to predict the possibility of fission of a nucleus.

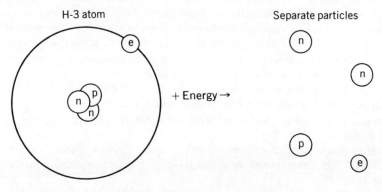

Fig. 2.8.   Dissociation of tritium.

We can speak of the binding energy associated with one particle such as a neutron. Suppose that M_1 is the mass of an atom and M_2 is its mass after absorbing a neutron. The binding energy of the neutron of mass m_n is then

$$B_n = M_1 + m_n - M_2.$$

2.7 SUMMARY

All material is composed of elements whose chemical interaction depends on the number of electrons (Z). Light is absorbed and emitted in the form of photons when atomic electrons jump between orbits. Isotopes of elements differ according to the number of nucleons in the nucleus (A). Nuclei are much smaller than atoms and contain most of the mass of the atom. The nucleons are bound together by a net force in which the nuclear attraction forces exceed the electrostatic repulsion forces. Energy must be supplied to dissociate a nucleus into its components.

2.8 PROBLEMS

2.1. Find the number of carbon ($^{12}_{6}C$) atoms in $1 \, cm^3$ of graphite, density $1.65 \, g/cm^3$.

2.2. Calculate the most probable speed of a "neutron gas" at temperature 20°C (293°K), noting that the mass of a neutron is $1.67 \times 10^{-27} \, kg$.

2.3. What frequency of light is emitted when an electron jumps into the smallest orbit of hydrogen, coming from a very large radius (assume infinity)?

2.4. Calculate the energy in electron-volts of the electron orbit in hydrogen for which $n = 3$, and find the radius in centimeters. How much energy would be needed to cause an electron to go from the innermost orbit to this one? If the electron jumped back, what frequency of light would be observed?

2.5. Sketch the atomic and nuclear structure of carbon-14, noting Z and A values and the numbers of electrons, protons, and neutrons.

2.6. What is the radius of the nucleus of uranium-238 viewed as a sphere? What is the area of the nucleus, seen from a distance as a circle?

2.7. Find the binding energy in MeV of ordinary helium ^4_2He, for which $M = 4.002603$.

2.8. How much energy (in MeV) would be required to completely dissociate the uranium-235 nucleus (atomic mass 235.043925) into its component protons and neutrons?

2.9. Find the mass density of the nucleus, the electrons, and the atom of U-235, assuming spherical shapes and the following data:

atomic radius	$1.7 \times 10^{-10} \, m$
nuclear radius	$8.6 \times 10^{-15} \, m$
electron radius	$2.8 \times 10^{-15} \, m$
mass of 1 amu	$1.66 \times 10^{-27} \, kg$
mass of electron	$9.11 \times 10^{-31} \, kg$

Discuss the results.

3

Radioactivity

Many naturally occurring and man-made isotopes have the property of radioactivity, which is the spontaneous disintegration (decay) of the nucleus with the emission of a particle. The process takes place in minerals of the ground, in fibers of plants, in tissues of animals, and in the air and water, all of which contain traces of radioactive elements.

3.1 NATURAL RADIOACTIVITY

Many heavy elements are radioactive. An example is the decay of the main isotope of uranium, in the reaction

$$^{238}_{92}U \rightarrow {}^{234}_{90}Th + {}^{4}_{2}He.$$

The particle released is the α (alpha) particle, which is merely the helium nucleus. The new isotope of thorium is also radioactive, according to

$$^{234}_{90}Th \rightarrow {}^{234}_{91}Pa + {}^{0}_{-1}e + \nu.$$

The three products are respectively the element protactinium, a β (beta) particle, which is merely an electron, and a neutrino symbolized by ν (nu). The latter is a neutral particle of zero rest mass that shares the reaction's energy release with the β particle. On the average, the neutrino carries $\frac{2}{3}$ of the energy, the electron, $\frac{1}{3}$. We note that the A value decreases by 4 and the Z value by 2 on emission of an α particle, while the A remains unchanged but Z increases by 1 on emission of a β particle. These two events are the start of a long sequence or "chain" of disintegrations that involve isotopes of the elements radium, polonium, and bismuth, eventually yielding the stable lead isotope $^{206}_{82}$Pb. Other chains found in nature start with $^{235}_{92}$U and $^{232}_{90}$Th. Hundreds of "artificial" radioisotopes have been produced by bombardment of nuclei by charged particles or neutrons, and by separation of the products of the fission process.

3.2 THE DECAY LAW

The rate at which a radioactive substance disintegrates (and thus the rate of release of particles) depends on the isotopic species, but there is a definite "decay law" that governs the process. In a given time period, say one second, each nucleus of a given isotopic species has the same chance of decay. If we were able to watch one nucleus, it might decay in the next instant, or a few days later, or even hundreds of years later. Such statistical behavior is described by means of the *half-life* t_H, which is the time required for half of the nuclei to decay. We should like to know how many nuclei of a radioactive species remain at any time. If we start at time

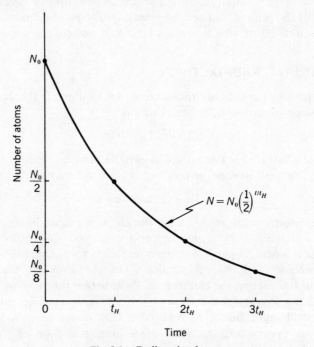

Fig. 3.1. Radioactive decay.

zero with N_0 nuclei, after a length of time t_H, there will be $N_0/2$; by the time $2t_H$ has elapsed, there will be $N_0/4$; etc. A graph of the number of nuclei as a function of time is shown in Fig. 3.1. For any time t on the curve, the ratio of the number of nuclei present to the initial number is given by

$$\frac{N}{N_0} = \left(\frac{1}{2}\right)^{t/t_H}.$$

Half-lives range from very small fractions of a second to billions of years, with each radioactive isotope having a definite half-life. Table 3.1 gives several examples of radioactive materials with their emissions, product isotopes, and half-lives. The β particle energies are maximum values; on the average the emitted betas have only one-third as much energy. Included in the table are both natural and man-made radioactive isotopes (also called radioisotopes). We note the special case of neutron decay according to

$$\text{neutron} \rightarrow \text{proton} + \text{electron}.$$

A free neutron has a half-life of 11.7 min. The conversion of a neutron into a proton can be regarded as the origin of beta emission in radioactive nuclei. Most of the radioisotopes in nature are heavy elements. One exception is potassium-40, half-life 1.28×10^9 yr, with abundance 0.118%

Table 3.1. Selected Radioactive Isotopes.†

Isotope	Half-life	Principal radiations (type, energy in MeV)
Neutron	10.6 m	β, 0.782
Hydrogen-3 (tritium)	12.33 y	β, 0.0186
Carbon-14	5730 y	β, 0.155
Sodium-24	15.03 h	β, 1.389; γ, 1.369, 2.754
Phosphorus-32	14.28 d	β, 1.711
Potassium-40	1.28×10^9 y	β, 1.325
Argon-41	1.83 h	β, 1.198; γ, 1.294
Cobalt-60	5.271 y	β, 0.318; γ, 1.173, 1.332
Krypton-85	10.7 y	β, 0.672; γ, 0.517
Strontium-90	28.8 y	β, 0.546
Iodine-131	8.040 d	β, 0.606
Xenon-135	9.10 h	β, 0.905
Cesium-137	30.17 y	β, 0.512
Radium-226	1.60×10^3 y	α, 4.78
Uranium-235	7.038×10^8 y	α, 4.40
Uranium-238	4.468×10^9 y	α, 4.20
Plutonium-239	2.41×10^4 y	α, 5.16

†Reference: *Table of Isotopes*, 7th edition, Ed. by C. Michael Lederer and Virginia S. Shirley. John Wiley & Sons, Inc., New York (1978).

in natural potassium. Others are carbon-14 and hydrogen-3 (tritium), which are produced continuously in small amounts by natural nuclear reactions. All three radioisotopes are found in plants and animals.

In addition to the radioisotopes that decay by beta or alpha emission, there is a large group of artificial isotopes that decay by the emission of a positron, which has the same mass as the electron and an equal but positive charge. An example is sodium-22, which decays with 2.6 yr half-life into a neon isotope as

$$\ce{^{22}_{11}Na} \rightarrow \ce{^{22}_{10}Ne} + \ce{^{0}_{+1}e}.$$

Whereas the electron (also called negatron) is a normal part of any atom, the positron is not, and eventually is annihilated by combination with an electron to produce photons, as will be discussed in Chapter 6.

A nucleus can get rid of excess internal energy by the emission of a gamma ray, but in an alternate process called internal conversion the energy is imparted directly to one of the atomic electrons, ejecting it from the atom. In an inverse process called K-capture, the nucleus spontaneously absorbs one of its own orbital electrons. Each of these processes is followed by the production of X-rays as the inner shell vacancy is filled.

The formula for N/N_0 is not very convenient for calculations except when t is some integer multiple of t_H. Defining the decay constant λ (lambda), as the chance of decay of a given nucleus each second, an equivalent *exponential relation*† for decay is

$$\frac{N}{N_0} = e^{-\lambda t}.$$

We find that $\lambda = 0.693/t_H$. To illustrate, let us calculate the ratio N/N_0 at the end of two years for cobalt-60, half-life 5.27 yr. This artificially produced radioisotope has many medical and industrial applications. The reaction is

$$\ce{^{60}_{27}Co} \rightarrow \ce{^{60}_{28}Ni} + \ce{^{0}_{-1}e} + 2\gamma,$$

---

†If λ is the chance one nucleus will decay in a second, then the chance in a time interval dt is λdt. For N nuclei, the change in number of nuclei is

$$dN = -\lambda N dt.$$

Integrating, and letting the number of nuclei at time zero be N_0 yields the formula quoted. Note that if

$$e^{-\lambda t} = \left(\frac{1}{2}\right)^{t/t_H},$$

then

$$\lambda t = \frac{t}{t_H} \log_e 2 \text{ or } \lambda = (\log_e 2)/t_H.$$

where the gamma ray energies are 1.17 and 1.33 MeV and the maximum beta energy is 0.318 MeV. Using the conversion 1 yr = 3.16 × 10^7 sec, t_H = 1.67 × 10^8 sec. Then λ = 0.693/(1.67 × 10^8) = 4.15 × 10^{-9} sec^{-1}, and since t is 6.32 × 10^7 sec, λt is 0.262 and $N/N_0 = e^{-0.262} = 0.77$.

The rate of release of radiation by a radioisotope is dependent on the *activity*, A, which is the number of disintegrations per second. Since the decay constant λ is the chance of decay each second, then with N nuclei present, the activity is

$$A = \lambda N.$$

Let us find the activity for a sample of Co-60 consisting of 1 microgram, 10^{16} atoms. Now $A = (10^{16})(4.15 \times 10^{-9}) = 4.15 \times 10^7$ disintegrations per second (d/sec).

A useful unit of activity is the curie (Ci), named for the French scientists who worked with radium. The curie is 3.7 × 10^{10} d/sec, which is an early measured value of the activity per gram of radium. Our cobalt sample has a "strength" of (4.15 × 10^7)/(3.7 × 10^{10}) = 0.0011 Ci or 1.1 mCi.

The half-life tells us how long it takes for half of the nuclei to decay, while a related quantity, the mean life, τ, (tau) is the average time elapsed for decay of an individual nucleus. It turns out that τ is $1/\lambda$ and thus equal to $t_H/0.693$. For Co-60, τ is 7.6 yr.

3.3 MEASUREMENT OF HALF-LIFE

Finding the half-life of an isotope provides part of its identification, needed for beneficial use or for protection against radiation hazard. Let us look at a method for measuring the half-life of a radioactive substance. As in Fig. 3.2, a detector that counts the number of particles striking it is placed near the source of radiation. From the number of counts observed in a known short time interval, the counting rate is computed. It is proportional to the rates of emission of particles or rays from the sample and thus to the activity A of the source. The process is repeated after an elapsed time for decay. The resulting values of activity are plotted on semilog graph paper as in Fig. 3.3, and a straight line drawn through the observed points. From any pairs of points on the line λ and $t_H = 0.693/\lambda$ can be calculated (see Problem 3.7). The technique may be applied to mixtures of two radioisotopes. After a long time has elapsed, only the isotope of longer half-life will contribute counts. By extending its graph linearly back in time, one can find the counts to be subtracted from the total to yield the counts from the isotope of shorter half-life.

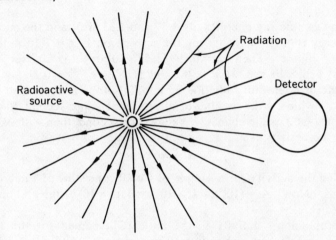

Fig. 3.2. Measurement of radiation from radioactive source.

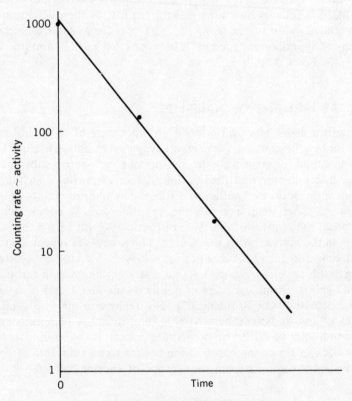

Fig. 3.3. Activity plot.

Activity plots cannot be used for a substance with very long half-life, e.g., strontium-90, 28.8 yr. The change in activity is almost zero over the span of time one is willing to devote to a measurement. However, if one knows the number of atoms present in the sample and measures the activity, the decay constant can be calculated from $\lambda = A/N$, from which t_H can be found.

The measurement of the activity of a radioactive substance is complicated by the presence of background radiation, which is due to cosmic rays from outside the earth or from the decay of minerals in materials of construction or in the earth. It is always necessary to measure the background counts and subtract them from those observed in the experiment.

3.4 SUMMARY

Many elements found in nature or man-made are radioactive, emitting alpha particles, beta particles, and gamma rays. The process is governed by an exponential relation, such that half of a sample decays in a time called the half-life t_H. Values of t_H range from fractions of a second to billions of years among the hundreds of radioisotopes known. Measurement of the activity, as the disintegration rate of a sample, yields half-life values, of importance in radiation use and protection.

3.5 PROBLEMS

3.1. Calculate the activity A for 1 g of radium-226 using the modern value of the half-life, and compare it with the definition of a curie.

3.2. The radioisotope sodium-24 ($^{24}_{11}$Na), half-life 15 hr, is used to measure the flow rate of salt water. By irradiation of stable $^{23}_{11}$Na with neutrons, suppose that we produce 5 micrograms of the isotope. How much do we have at the end of 24 hr?

3.3. What was the initial and final activity in disintegrations per second and in curies for the sample of Na-24 in Problem 3.2?

3.4. The isotope uranium-238 ($^{238}_{92}$U) decays successively to form $^{234}_{90}$Th, $^{234}_{91}$Pa, $^{234}_{92}$U, $^{230}_{90}$Th, finally becoming radium-226 ($^{226}_{88}$Ra). What particles are emitted in each of these five steps? Draw a graph of this chain, using A and Z values on the horizontal and vertical axes, respectively.

3.5. A capsule of cesium-137, half-life 30.2 yr, is used to check the accuracy of detectors of radioactivity in air and water. Draw a graph on semilog paper of the activity over a 10-yr period of time, assuming the initial strength is 1 mCi. Explain the results.

3.6. A typical person has 140 g of potassium in his body. Of this a fraction 1.18×10^{-3} is K-40, half-life 1.28×10^{9} yr. How many disintegrations occur each second in the body?

3.7. (a) Noting that the activity of a radioactive substance is $A = \lambda N_0 e^{-\lambda t}$, verify that the graph of counting rate versus time on semilog paper is a straight line and show that

$$\lambda = \frac{\log_e (C_1/C_2)}{t_2 - t_1}$$

where points 1 and 2 are any pair on the curve.

(b) Using the following data, deduce the half-life of an "unknown," and suggest what isotope it is.

Time (sec)	Counting Rate (per sec)
0	200
1000	182
2000	162
3000	144
4000	131

3.8. By chemical means, we deposit 10^{-8} moles of a radioisotope on a surface and measure the activity to be 82,000 d/sec. What is the half-life of the substance and what element is it (see Table 3.1)?

4

Nuclear Reactions

Nuclear reactions are processes in which some change in the character of a nucleus takes place, either spontaneously as in radioactivity, or as the result of bombardment by a particle or ray.

4.1 ANALOGIES BETWEEN NUCLEAR AND CHEMICAL REACTIONS

There are many similarities between the two types of reactions, even though they occur in quite different energy regions and with different mechanisms. Individual particles are involved in each case—nucleons in nuclei and atoms in molecules, respectively. Conservation laws apply to each—that of charge, number of particles involved, and mass-energy. Reactions are similar in appearance. To illustrate, let us first write the simple equation describing the formation of water

$$2\,H + O = H_2O.$$

The numbers of atoms of hydrogen and oxygen are the same on both sides; there is a valence balance, with $+1$ for hydrogen and -2 for oxygen; energy is released in the process, in this case 2.4 eV per molecule of water. Compare this with the reaction of a neutron with a proton, the nucleus of hydrogen, to form a deuteron, the nucleus of deuterium.

$$_0^1n + {}_1^1H \rightarrow {}_1^2H + \gamma.$$

The total number of nucleons is the same on both sides; the total nuclear charge is balanced; energy is released here in the form of a gamma ray, of about 2.2 MeV. We note that the energy release in this nuclear reaction is about a million times as great as for the chemical or atomic reaction.

A very large number of possible chemical reactions can take place because there are more than 100 known elements. Similarly, there are many nuclear reactions, involving about 2000 known isotopes, and several particles that can serve to induce reactions or be products of reaction—photons, electrons, protons, neutrons, alpha particles, deuterons, and heavier charged particles.

4.2   NUCLEAR TRANSMUTATION

The first example of artificial conversion of one element into another, a process of transmutation, was discovered by Rutherford in 1919. The reaction involved the bombardment of an isotope of nitrogen by alpha particles (nuclei of helium) from a radioactive source. The reaction products were an isotope of oxygen and a proton. The equation for the process is

$$\,^4_2\text{He} + \,^{14}_7\text{N} \rightarrow \,^{17}_8\text{O} + \,^1_1\text{H}.$$

We note that on both sides of the equation the sum of the A values is the same (18) and the sum of the Z values is the same (9). Figure 4.1 shows

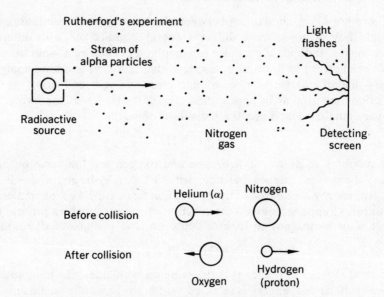

Fig. 4.1.   Transmutation by nuclear reaction.

the reaction schematically. It is difficult for the alpha particle to enter the nitrogen nucleus because of the electrostatic repulsion of the two positively charged interacting nuclei. Thus the alpha particle must have several million-electron-volts of energy. Nuclear transmutations can also be induced by charged particles accelerated electrically to high speed. The first one discovered was

$$_1^1H + _3^7Li \rightarrow 2\,_2^4He.$$

Another reaction yielding a radioactive isotope of nitrogen is

$$_1^1H + _6^{12}C \rightarrow _7^{13}N + \gamma.$$

Nitrogen-13 emits a positron, the positive counterpart of the electron, with a half-life of 10 min.

4.3 NEUTRON REACTIONS

In contrast with charged particles, the neutron as a neutral particle need not have a high energy to penetrate the nucleus. Neutrons are thus especially effective as projectiles for inducing nuclear reactions. One of practically zero energy can be captured by hydrogen in the example described previously. That reaction yielded the stable isotope deuterium, normally present in nature. A radioactive isotope is produced by the reaction

$$_0^1n + _{27}^{59}Co \rightarrow _{27}^{60}Co + \gamma.$$

The cobalt-59 isotope has been changed into one of higher atomic mass, cobalt-60, which subsequently decays as discussed in Chapter 3. Note that energy is released instantly in the form of a capture gamma ray in the reaction that produces the radioisotope, and that energy is released much later in the form of beta particles, neutrinos, and decay gamma rays. Another example is neutron capture in cadmium, often used in control rods of a nuclear reactor, according to

$$_0^1n + _{48}^{113}Cd \rightarrow _{48}^{114}Cd + \gamma.$$

A reaction that may some day be employed to produce tritium, one of the fuels for a controlled thermonuclear reactor, is

$$_0^1n + _3^6Li \rightarrow _1^3H + _2^4He.$$

The alchemist's dream was realized when the reaction to produce gold from mercury was first performed,

$$_0^1n + _{80}^{198}Hg \rightarrow _{79}^{198}Au + _1^1H.$$

4.4 ENERGY BALANCES

The conservation of mass-energy is a firm requirement for any valid nuclear reaction. Recall from Chapter 1 that the total energy is made up of the rest energy (corresponding to a particle's mass at rest), and the kinetic energy of motion. The total energy of the reactants must equal that of the products. Strictly speaking, we should account for all of the effects of special relativity, including the mass increase with particle speed. Fortunately, however, for reactions of particles that come together and leave each other at low speeds we can use the classical formula $E_k = \frac{1}{2} m_0 v^2$ for the kinetic energy and the inherent energy associated with the rest mass $E_0 = m_0 c^2$.

Let us calculate the energy release from the neutron–proton reaction, assuming that the neutron is so slow that its kinetic energy can be neglected. Write a balance statement

mass of neutron + mass of hydrogen atom =
mass of deuterium atom + kinetic energy of products.

The kinetic energy is that of the $^2_1$H atom plus that of the gamma ray, the latter having no rest mass. We insert the accurately known masses of the particles,
$$1.008665 + 1.007825 = 2.014102 + E_k.$$

Solving, E_k is 0.002378 amu, and since 1 amu = 931 MeV, the energy release is 2.22 MeV.

We can further illustrate the energy balance idea by finding the energy of the two alpha particles released in the reaction of a proton with lithium-7. Suppose that the proton has a kinetic energy of 2 MeV which corresponds to 2/931=0.002148 amu and that the target nucleus is at rest. The energy balance statement is

kinetic energy of hydrogen + mass of hydrogen + mass of lithium =
mass of helium + kinetic energy of helium
$$0.002148 + 1.007825 + 7.016004 = 2(4.002603) + E_k$$

Then $E_k = 0.02077$ amu = 19.3 MeV. This energy is shared by the two alpha particles.

4.5 MOMENTUM CONSERVATION

The calculations just completed tell us the total kinetic energy of the product particles but do not say how much each has or what their speeds are. To find the kinetic energies of each of the particles we must apply the principle of conservation of momentum. Recall that the linear momentum p of a material particle of mass m and speed v is $p = mv$. This relation is correct in both the classical and relativistic senses. The total momentum of the interacting particles before and after the reaction is the same.

For our problem of a very slow neutron striking a hydrogen atom at rest, we can assume the initial momentum is zero. If it is to be zero finally, the ^2_1H and γ-ray must fly apart with equal magnitudes of momentum $p_d = p_\gamma$. The momentum of a gamma ray having the speed of light c may be written $p_\gamma = mc$ if we regard the mass as an *effective* value, related to the gamma energy E_γ by Einstein's formula $E = mc^2$. Thus

$$p_\gamma = \frac{E_\gamma}{c}.$$

Most of the 2.22 MeV energy release of the neutron capture reaction goes to the gamma ray, as shown in Problem 4.5. Assuming that to be correct, we can estimate the effective mass of this gamma ray. It is close to 0.00238 amu, which is very small compared with 2.014 amu for the deuterium. Then from the momentum balance, we see that the speed of recoil of the deuterium is very much smaller than the speed of light.

The calculation of the energies of the two alpha particles is a little complicated even for the case in which they separate along the same line that the proton entered. If we let m be the alpha particle mass and v_1 and v_2 be their speeds, with p_H the proton momentum, we must solve the two equations

$$mv_1 - mv_2 = p_H,$$
$$\tfrac{1}{2}mv_1{}^2 + \tfrac{1}{2}mv_2{}^2 = E_k.$$

4.6 GENERAL CONCEPTS OF NUCLEAR REACTIONS

A systematic description of nuclear reactions has been developed to account for experimental observations. Algebraic symbols are used to represent the particles involved in a sequence of events, as follows. Suppose that one particle labeled a strikes a nucleus X to produce a "compound nucleus" C^*, where the asterisk implies that the nucleus contains extra internal energy of motion of the nucleons, called excitation energy. The compound nucleus then disintegrates to release the particle b and a residual nucleus Y. In terms of reaction equations, this may be written

$$a + X \rightarrow C^*,$$
$$C^* \rightarrow Y + b,$$

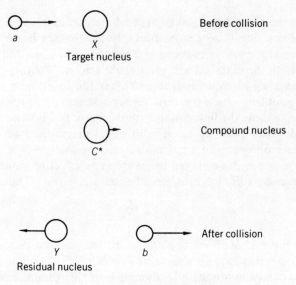

Before collision

Target nucleus

Compound nucleus

After collision

Residual nucleus

Fig. 4.2. General nuclear reaction.

with the net effect being

$$a + X = Y + b.$$

Figure 4.2 shows the event schematically. This reaction can be abbreviated as

$$X(a, b)Y,$$

which implies that particle a comes in and particle b goes out. The symbols a and b may stand for the neutron (n), alpha particle (α), deuteron (d), gamma ray (γ), proton (p), triton (t), the nucleus of tritium, and so on. In the previous sections, we have given examples of these reactions: (n, γ), (α, p), (p, α), (p, γ), (n, t), and (n, p).

For reactions at low energy, less than 10 MeV, the same compound nucleus could be formed by several different pairs of interacting nuclei, and could decay into several different pairs of final products. Consider Rutherford's transmutation reaction. The compound nucleus C^* in this case is the isotope $^{18}_{9}F^*$. It could have been formed in other ways, for instance by the combination of $^{1}_{1}H$ and $^{17}_{8}O$ or $^{1}_{0}n$ and $^{17}_{9}F$; it could disintegrate into $^{16}_{8}O$ and $^{2}_{1}H$ or $^{18}_{9}F$ and a gamma ray.

Depending on the binding energies of the nuclei, some of these reactions will release energy, others will require energy to be supplied. In the latter case, no nuclear reaction will take place if the energy of the incident particle is too low—the projectile will merely be scattered by the target. If the masses of all the particles involved are accurately known, the consequences of bombardment of a nucleus by any particle of any energy can be deduced, using the laws of conservation of energy and momentum.

4.7 SUMMARY

Chemical and nuclear reactions have several similarities in the form of equations and the requirements on conservation of particles and charge. The bombardment of nuclei by charged particles or neutrons produces new nuclei and particles. The final energies are found by taking account of mass difference, and the final speeds are obtained by applying the law of momentum conservation. In general, a compound nucleus having excitation energy is an intermediate step in a nuclear reaction.

4.8 PROBLEMS

4.1. The energy of formation of water from its constituent gases is quoted to be 54,500 cal/mole. Verify that this corresponds to 2.4 eV per molecule of H_2O.

4.2. Complete the following nuclear reaction equations:

$$_{0}^{1}n + {}_{7}^{14}N \rightarrow \{\ \}(\quad) + {}_{1}^{1}H,$$
$$_{1}^{2}H + {}_{4}^{9}Be \rightarrow \{\ \}(\quad) + {}_{0}^{1}n.$$

4.3. Using the accurate atomic masses listed below, find the minimum amount of energy an alpha particle must have to cause the transmutation of nitrogen to oxygen. (${}_{7}^{14}N$ 14.003074, ${}_{2}^{4}He$ 4.002603, ${}_{8}^{17}O$ 16.999131, ${}_{1}^{1}H$ 1.007825.)

4.4. Find the energy release in the reaction ${}_{3}^{6}Li$ (n, α) ${}_{1}^{3}H$, noting the masses ${}_{0}^{1}n$ 1.008665, ${}_{1}^{3}H$ 3.016049, ${}_{2}^{4}He$ 4.002603, and ${}_{3}^{6}Li$ 6.015123.

4.5. A slow neutron of mass 1.008665 amu is caught by the nucleus of a hydrogen atom of mass 1.007825 and the final products are a deuterium atom of mass 2.014102 and a gamma ray. The energy released is 2.22 MeV. If the gamma ray is assumed to have almost all of this energy, what is its effective mass in kg? What is the speed of the ${}_{1}^{2}H$ particle in m/sec, using equality of momenta on separation? What is the recoil energy of ${}_{1}^{2}H$ in MeV? How does this compare with the total energy released? Was the assumption about the gamma ray reasonable?

4.6. Calculate the speeds and energies of the individual alpha particles in the reaction ${}_{1}^{1}H + {}_{3}^{7}Li \rightarrow 2 {}_{2}^{4}He$, assuming that they separate along the line of proton motion. Note that the mass of the lithium-7 atom is 7.016004.

4.7. Calculate the energy release in the reaction

$$^{13}_{7}N \rightarrow \quad ^{13}_{6}C + \quad ^{0}_{+1}e$$

where atomic masses are $^{13}_{7}N$ 13.005739, $^{13}_{6}C$ 13.003355, and the masses of the positron and electron are 0.000549. *Note:* In reactions involving positrons it is necessary to use masses of nuclei rather than atoms. Explain.

5

Reaction Rates

The chance of interaction of any two particles obviously depends on the nature of the force between them. The most familiar force is that of electrostatics, described by Coulomb's relation $F \sim q_1 q_2 / r^2$, where the q's are the charges carried by the objects and r is the distance of separation of their centers. Two charged particles, such as electrons or protons, will influence each other somewhat, no matter how far they are apart. On the other hand, two atoms, which are each neutral because of the equality of electronic and nuclear charge, will not interact appreciably until they get close to each other (around 10^{-8} cm). The special force between nucleons is similarly limited to small distances of separation (around 10^{-13} cm).

5.1 CROSS SECTIONS FOR PARTICLE INTERACTION

It is natural to think of a particle as having a definite size and, if viewed as a sphere, as having a radius. The question arises as to how such a small radius can be found. For objects of this size, the use of ordinary measuring devices such as a ruler is evidently impossible. Some form of microscope is required, using electromagnetic radiation—X-rays or gamma rays—or material particles—electrons, protons, or neutrons. However, the chance of interaction depends on the properties of both the particle used as probe and the particle under scrutiny. The radius deduced thus depends on the method of measurement. We must be satisfied with information on the apparent radius or the apparent cross-sectional area of target, each of which depends on the particular interaction. This leads to the concept of cross section, as a measure of the chance of collision between any two particles, whether they be atoms, nuclei, or photons.

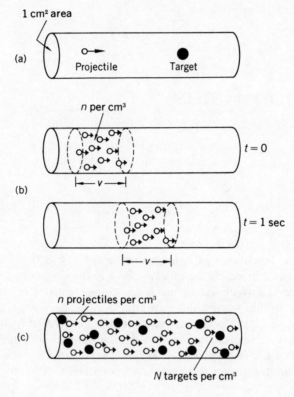

Fig. 5.1. Particle collisions.

We can perform a set of imaginary experiments that will clarify the idea of cross section. Picture, as in Fig. 5.1a, a tube of end area $1\,cm^2$ containing only one target particle. A single projectile is injected parallel to the tube axis, but its exact location is not specified. It is clear that the chance of collision, labeled σ (sigma) and called the *microscopic cross section*, is the ratio of the target area to the area of the tube, which is 1.

Now let us inject a continuous stream of particles of speed v into the empty tube (see Fig. 5.1b). In a time of one second, each of the particles has moved along a distance v cm. All of them in a column of volume $(1\,cm^2)(v\ cm) = v\ cm^3$ will sweep past a point at which we watch each second. If there are n particles per cubic centimeter, then the number per unit time that cross any unit area perpendicular to the stream direction is nv, called the *current density*.

Finally, Fig. 5.1c, we fill each unit volume of the tube with N targets, each of area σ as seen by incoming projectiles (we presume that the targets do not "shadow" each other). If we focus attention on a unit volume, there is a total target area of $N\sigma$. Again, we inject the stream of projectiles. In a time of one second, the number of them that pass through the target volume is nv; and since the chance of collision of each with one target atom is σ, the number of collisions is $nvN\sigma$. We can thus define the reaction rate per unit volume,

$$R = nvN\sigma.$$

We let the current density nv be abbreviated by j and let the product $N\sigma$ be labeled Σ (capital sigma), the *macroscopic cross section*, referring to the large-scale properties of the medium. Then the reaction rate per cubic centimeter is simply $R = j\Sigma$. We can easily check that the units of j are $\text{cm}^{-2}\,\text{sec}^{-1}$ and those of Σ are cm^{-1}, so that the units of R are $\text{cm}^{-3}\,\text{sec}^{-1}$.

In a different experiment, we release particles in a medium and allow them to make many collisions with those in the material. In a short time, the directions of motion are random, as sketched in Fig. 5.2. We shall look only at particles of the same speed v, of which there are n per unit volume. The product nv in this situation is no longer called current density, but is given a different name, the flux, symbolized by ϕ (phi). If we place a unit area anywhere in the region, there will be flows of particles across it each second from both directions, but it is clear that the current densities will now be less than nv. It turns out that they are each

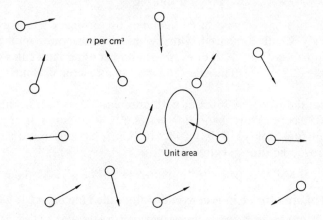

Fig. 5.2. Particles in random motion.

$nv/4$, and the total current density is $nv/2$. The rate of reaction of particles with those in the medium can be found by adding up the effects of individual projectiles. Each behaves the same way in interacting with the targets, regardless of direction of motion. The reaction rate is again $nvN\sigma$ or, for this random motion, $R = \phi\Sigma$.

When a particle such as a neutron collides with a target nucleus, there is a certain chance of each of several reactions. The simplest is elastic scattering, in which the neutron is visualized as bouncing off the nucleus and moving in a new direction with a change in energy. Such a collision, governed by classical physics, is predominant in light elements. In the inelastic scattering collision, an important process for fast neutrons in heavy elements, the neutron becomes a part of the nucleus, its energy provides excitation, and a neutron is released. The cross section σ_s is the chance of a collision that results in neutron scattering. The neutron may instead be absorbed by the nucleus, with cross section σ_a. Since σ_a and σ_s are chances of reaction, their sum is the chance for collision or total cross section $\sigma = \sigma_a + \sigma_s$.

Let us illustrate these ideas by some calculations. In a typical nuclear reactor used for training and research in universities, a large number of neutrons will be present with energies near 0.0253 eV. This energy corresponds to a most probable speed of 2200 m/sec for the neutrons viewed as a gas at room temperature, $293°$ absolute. Suppose that the flux of such neutrons is 2×10^{12} cm^{-2}-sec^{-1}. The number density is then

$$n = \frac{\phi}{v} = \frac{2 \times 10^{12}\,\text{cm}^{-2}\text{-sec}^{-1}}{2.2 \times 10^5\,\text{cm/sec}} = 9 \times 10^6\,\text{cm}^{-3}.$$

Although this is a very large number by ordinary standards, it is exceedingly small compared with the number of water molecules per cubic centimeter (3.3×10^{22}) or even the number of air molecules per cubic centimeter (2.7×10^{19}). The "neutron gas" in a reactor is almost a perfect vacuum.

Now let the neutrons interact with uranium-235 as fuel in the reactor. The cross section for absorption σ_a is 678×10^{-24} cm^2. If the number density of fuel atoms is $N = 0.048 \times 10^{24}$ cm^{-3}, as in uranium metal, then the macroscopic cross section is

$$\Sigma_a = N\sigma_a = (0.048 \times 10^{24}\,\text{cm}^{-3})(678 \times 10^{-24}\,\text{cm}^2) = 32.5\,\text{cm}^{-1}.$$

The unit of area 10^{-24} cm^2 is conventionally called the *barn*.† If we express

---

†As the story goes, an early experimenter observed that the cross section for U-235 was "as big as a barn."

the number of targets per cubic centimeter in units of 10^{24} and the microscopic cross section in barns, then $\Sigma_a = (0.048)(678) = 32.5$ as above. With a neutron flux $\phi = 3 \times 10^{13} \text{cm}^{-2}\text{-sec}^{-1}$, the reaction rate for absorption is

$$R = \phi \Sigma_a = (3 \times 10^{13} \text{cm}^{-2}\text{-sec}^{-1})(32.5 \text{ cm}^{-1}) = 9.75 \times 10^{14} \text{ cm}^{-3}\text{-sec}^{-1}.$$

This is also the rate at which uranium-235 nuclei are consumed.

The average energy of neutrons in a nuclear reactor used for electrical power generation is about 0.1 eV, almost four times the value used in our example. The effects of the high temperature of the medium (about 600°F) and of neutron absorption give rise to this higher value.

5.2 PARTICLE ATTENUATION

Visualize an experiment in which a stream of particles of common speed and direction is allowed to strike the plane surface of a substance as in Fig. 5.3. Collisions with the target atoms in the material will continually remove projectiles from the stream, which will thus diminish in strength with distance, a process we label attenuation. If the current density incident on the substance at position $z = 0$ is labeled j_0, the

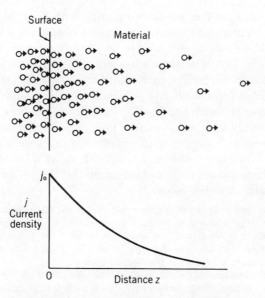

Fig. 5.3. Neutron penetration and attenuation.

current of those not having made any collision on penetrating to a depth z is given by[†]

$$j = j_0 e^{-\Sigma z},$$

where Σ is the macroscopic cross section. The similarity in form to the exponential for radioactive decay is noted, and one can deduce by analogy that the half-thickness, the distance required to reduce j to half its initial value, is $z_H = 0.693/\Sigma$. Another more frequently used quantity is the mean free path λ, the average distance a particle goes before making a collision. By analogy with the mean life for radioactivity, we can write[‡]

$$\lambda = \frac{1}{\Sigma}.$$

This relation is applicable as well to particles moving randomly in a medium. Consider a particle that has just made a collision and moves off in some direction. On the average, it will go a distance λ through the array of targets before colliding again. For example, we can find the mean free path for 1 eV neutrons in water, assuming that scattering by hydrogen with cross section 20 barns is the dominant process. Now the number of hydrogen atoms is $N_H = 0.0668 \times 10^{24}\,\text{cm}^{-3}$, σ_s is $20 \times 10^{-24}\,\text{cm}^2$, and $\Sigma_s = 1.34\,\text{cm}^{-1}$. Thus the mean free path for scattering λ_s, is around 0.75 cm.

The cross sections for *atoms* interacting with their own kind at the energies corresponding to room temperature conditions are of the order of $10^{-15}\,\text{cm}^2$. If we equate this area to πr^2, the calculated radii are of the order of 10^{-8} cm. This is in rough agreement with the theoretical radius of electron motion in the hydrogen atom 0.53×10^{-8} cm. On the other hand, the cross sections for *neutrons* interacting with nuclei by *scattering* collisions, those in which the neutron is deflected in direction and loses energy, are usually very much smaller than those for atoms. For the case of 1 eV neutrons in hydrogen with a scattering cross section of 20 barns, i.e., $20 \times 10^{-24}\,\text{cm}^2$, one deduces a radius of about 2.5×10^{-12} cm. These results correspond to our earlier observation that the nucleus is thousands of times smaller than the atom.

[†]The derivation proceeds as follows. In a slab of material of unit area and infinitesimal thickness dz, the target area will be $N\sigma\,dz$. If the current at z is j, the number of collisions per second in the slab is $jN\sigma\,dz$, and thus the change in j on crossing the layer is $dj = -j\Sigma\,dz$ where the reduction is indicated by the negative sign. By analogy with the solution of the radioactive decay law, we can write the formula cited.

[‡]This relation can be derived directly by use of the definition of an average as the sum of the distances the particles travel divided by the total number of particles. Using integrals, this is $\bar{z} = \int z\,dj / \int dj$.

5.3 NEUTRON CROSS SECTIONS

The cross section for neutron *absorption* in materials depends greatly on the isotope bombarded and on the neutron energy. For consistent comparison and use, the cross section is often cited at 0.0253 eV, corresponding to neutron speed 2200 m/sec. Values for absorption cross sections for a number of isotopes at that energy are listed in order of increasing size in Table 5.1. The dependence of absorption cross section on energy is of two types, one called $1/v$, in which σ_a varies inversely with neutron speed, the other called resonance, where there is a very strong absorption at certain neutron energies. Many materials exhibit both variations. Figures 5.4 and 5.5 show the cross sections for boron and natural uranium. The use of the logarithmic plot enables one to display the large range of cross section over the large range of energy of interest. Neutron scattering cross sections are more nearly the same for all elements and have less variation with neutron energy. Figure 5.6 shows the trend of σ_s for hydrogen as in water. Over a large range of neutron energy the scattering cross section is nearly constant, dropping off in the million-electron-volt region. This high energy range is of special interest since neutrons produced by the fission process have such energy values.

Table 5.1. Selected Thermal Neutron Absorption Cross Sections (in Order of Increasing Size)†

Isotope	σ_a (barns)
$^{4}_{2}\text{He}$	$\simeq 0$
$^{16}_{8}\text{O}$	1.78×10^{-4}
$^{2}_{1}\text{H}$	5.2×10^{-4}
$^{12}_{6}\text{C}$	0.0034
$^{}_{40}\text{Zr}$	0.183
$^{1}_{1}\text{H}$	0.332
$^{238}_{92}\text{U}$	2.7
$^{235}_{92}\text{U}$	678
$^{239}_{94}\text{Pu}$	1013
$^{10}_{5}\text{B}$	3838.0
$^{135}_{54}\text{Xe}$	2.6×10^{6}

†Reference: *Table of Isotopes,* 7th edition, Ed. by C.M. Lederer and V. S. Shirley. John Wiley (1978). Based on information from N. Holden.

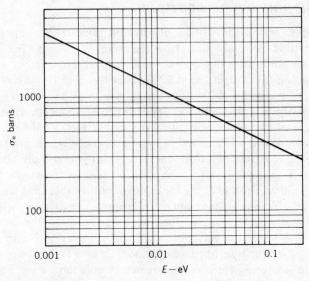

Fig. 5.4.   Absorption cross section for elemental boron.

5.4   NEUTRON SLOWING

When fast neutrons, those of energy of the order of 2 MeV, are introduced into a medium, a sequence of collisions with nuclei takes place. The neutrons are deflected in direction on each collision, they lose energy, and they tend to migrate away from their origin. Each neutron has a unique history, and it is impractical to keep track of all of them. Instead, we seek to deduce average behavior. First, we note that the elastic scattering of a neutron with a nucleus of mass number A causes a reduction in neutron energy from E_0 to E and a change of direction through an angle θ (theta), as sketched in Fig. 5.7. The length of arrows indicates the speeds of the particles. This example shown is but one of a great variety of possible results of scattering collisions. For each final energy there is a unique angle of scattering, and vice versa, but the occurrence of a particular E and θ pair depends on chance. The neutron may bounce directly backward, $\theta = 180°$, dropping down to a minimum energy αE_0, where $\alpha = (A - 1)^2/(A + 1)^2$, or it may be undeflected, $\theta = 0°$, and retain its initial energy E_0, or it may be scattered through any other angle, with corresponding energy loss.

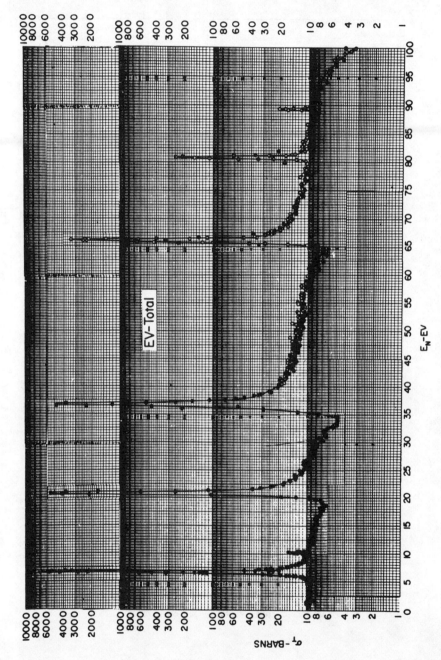

Fig. 5.5.   Cross section for natural uranium.

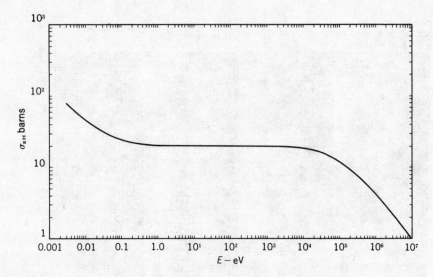

Fig. 5.6. Scattering cross section for hydrogen.

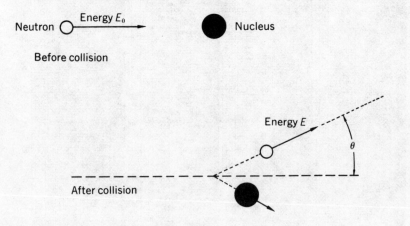

Fig. 5.7. Neutron scattering and energy loss.

The average elastic scattering collision is described by two quantities that depend only on the nucleus, not on the neutron energy. The first is $\overline{\cos\,\theta}$, the average of the cosines of the angles of scattering, given by

$$\overline{\cos\,\theta} = \frac{2}{3A}.$$

For hydrogen, it is $\frac{2}{3}$, meaning that the neutron tends to be scattered in the forward direction; for a very heavy nucleus such as uranium, it is near zero, meaning that the scattering is almost equally likely in each direction. Forward scattering results in an enhanced migration of neutrons from their point of appearance in a medium. Their free paths are effectively longer, and it is conventional to use the "transport mean free path" $\lambda_t = \lambda_s/(1 - \overline{\cos\,\theta})$ instead of λ_s to account for the effect. We note that λ_t is always the larger. Consider slow neutrons in carbon, for which $\sigma_s = 4.8$ barns and $N = 0.083$, so that $\Sigma_s = 0.4\,\mathrm{cm}^{-1}$ and $\lambda_s = 2.5$ cm. Now $\overline{\cos\,\theta} = 2/(3)(12) = 0.056$, $1 - \overline{\cos\,\theta} = 0.944$, and $\lambda_t = 2.5/0.944 = 2.7$ cm.

The second quantity that describes the average collision is ξ (xi), the average change in the natural logarithm of the energy, given by

$$\xi = 1 + \frac{\alpha\,\ln\,\alpha}{1 - \alpha}.$$

For hydrogen, it is exactly 1, the largest possible value, meaning that hydrogen is a good "moderator" for neutrons, its nuclei permitting the greatest neutron energy loss; for a heavy element it is $\xi \simeq 2/(A + \frac{2}{3})$ which is much smaller than 1, e.g., for carbon, $A = 12$, it is 0.16.

To find how many collisions C are required to slow neutrons from one energy to another, we merely divide the total change in $\ln E$ by ξ, the average per collision. In going from the fission energy 2×10^6 eV to the thermal energy 0.025 eV, the total change is $\ln(2 \times 10^6) - \ln(0.025) = \ln(8 \times 10^7) = 18.2$. Then $C = 18.2/\xi$. For example in hydrogen, $\xi = 1$, C is 18, while in carbon $\xi = 0.16$, C is 114. Again, we see the virtue of hydrogen as a moderator. The fact that hydrogen has a scattering cross section of 20 barns over a wide range while carbon has a σ_s of only 4.8 barns implies that collisions are more frequent and the slowing takes place in a smaller region. The only disadvantage is that hydrogen has a larger thermal neutron absorption cross section, 0.332 barns versus 0.0034 for carbon.

The movement of individual neutrons through a moderator during slowing consists of free flights, interrupted frequently by collisions that cause energy loss. Picture, as in Fig. 5.8, a fast neutron starting at a point,

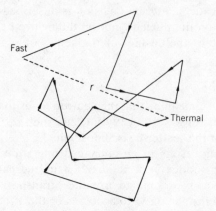

Fig. 5.8. Neutron migration during slowing.

and migrating outward. At some distance r away, it arrives at the thermal energy. Other neutrons become thermal at different distances, depending on their particular histories. If we were to measure all of their r values and form the average of r^2, the result would be $\overline{r^2} = 6\tau$, where τ (tau) is called the "age" of the neutron. Approximate values of the age for various reactor materials, as obtained from experiment are listed below

Moderator	τ, age to thermal (cm^2)
H_2O	26
D_2O	125
C	364

We thus note that water is a much better agent for neutron slowing than is graphite.

5.5 NEUTRON DIFFUSION

As neutrons slow into the energy region that is comparable to thermal agitation of the moderator atoms, they may either lose or gain energy on collision. Members of a group of neutrons have various speeds at any instant and thus the group behaves as a gas, with temperature T that is close to that of the medium in which they are found. Thus if the moderator is at room temperature 20°C or 293°K, the most likely neutron speed is around 2200 m/sec, corresponding to a kinetic energy of 0.0253 eV. To a first approximation the neutrons have a maxwellian distribution comparable to that of a gas, as was shown in Fig. 2.1.

The process of diffusion of gas molecules is familiar to us. If a bottle of perfume is opened, the scent is quickly observed, as the molecules of the substance migrate away from the source. Since neutrons in large numbers behave as a gas, the descriptions of gas diffusion may be applied. The flow of neutrons through space at a location is proportional to the way the concentration of neutrons varies, in particular to the negative of the slope of the neutron number density. We can guess that the larger the neutron speed v and the larger the transport mean free path λ_t, the more neutron flow will take place. Theory and measurement show that if n varies in the z-direction, the net flow of neutrons across a unit area each second, the net current density, is

$$j = \frac{-\lambda_t v}{3} \frac{dn}{dz}.$$

This is called Fick's law of diffusion, derived long ago for the description of gases. It applies if absorption is small compared with scattering. In terms of the flux $\phi = nv$ and the *diffusion coefficient* $D = \lambda_t/3$, this may be written compactly $j = -D\phi'$, where ϕ' is the slope of the neutron flux.

5.6 SUMMARY

The cross section for interaction of neutrons with nuclei is a measure of chance of collision. The reaction rate per cubic centimeter is mutually dependent on current density (j) and macroscopic cross section (Σ). The stream formed by uncollided particles is reduced exponentially as it passes through a medium. Neutron absorption cross sections vary greatly with target isotope and with neutron energy, while scattering cross sections are almost constant. Neutrons are slowed readily by collisions with light nuclei and migrate from their point of origin. On reaching thermal energy they continue to disperse, and the net flow is proportional to the negative of the slope of the flux.

5.7 PROBLEMS

5.1. Calculate the macroscopic cross section for scattering of 1 eV neutrons in water, using N for water as 0.0334×10^{24} cm^{-3} and cross sections 20 barns for hydrogen and 3.8 barns for oxygen. Find the mean free path λ_s.

5.2. Find the speed v and the number density of neutrons of energy 1.5 MeV in a flux 7×10^{13} cm^{-2}-sec^{-1}.

5.3. Compute the flux, macroscopic cross section and reaction rate for the following data: $n = 2 \times 10^5$ cm^{-3}, $v = 3 \times 10^8$ cm/sec, $N = 0.04 \times 10^{24}$ cm^{-3}, $\sigma = 0.5 \times 10^{-24}$ cm^2.

5.4. What are the values of the average logarithmic energy change ξ and the average cosine of the scattering angle $\overline{\cos} \, \theta$ for neutrons in beryllium, $A = 9$? How many collisions are needed to slow neutrons from 2 MeV to 0.025 eV in Be-9? What is the value of the diffusion coefficient D for 0.025 eV neutrons if Σ_s is 0.90 cm^{-1}?

5.5. (a) Verify that neutrons of speed 2200 m/sec have an energy of 0.0253 eV.
(b) If the neutron absorption cross section of boron at 0.0253 eV is 760 barns, what would it be at 0.1 eV? Does this result agree with that shown in Fig. 5.4?

5.6. Calculate the rate of consumption of U-235 and U-238 in a flux of 2.5×10^{13} cm^{-2}-sec^{-1} if U atom number density is of U atoms is 0.0223×10^{24} cm^{-3}, the atom number fractions of the two isotopes are 0.0072 and 0.9928, and cross sections are 678 barns and 2.70 barns, respectively. Comment on the results.

5.7. How many atoms of boron-10 per atom of carbon-12 would result in an increase of 50% in the macroscopic cross section of graphite? How many ^{10}B atoms would there then be per million ^{12}C atoms?

5.8. Calculate the absorption cross section of the element zirconium using the following isotopic data (mass number, fractional abundance, and cross section) 90, 0.515, 0.03; 91, 0.112, 1.1; 92, 0.171, 0.2; 94, 0.174, 0.055; 96, 0.028, 0.020. Compare with the figure given in Table 5.1.

6

Radiation and Materials

The word "radiation" will be taken to embrace all particles, whether they are of material or electromagnetic origin. We include those produced by both atomic and nuclear processes and those resulting from electrical acceleration, noting that there is no essential difference between X-rays from atomic collisions and gamma rays from nuclear decay; protons can come from a particle accelerator, from cosmic rays, or from a nuclear reaction in a reactor. The word "materials" will refer to bulk matter, whether of mineral or biological origin, as well as the particles of which the matter is composed, including molecules, atoms, electrons, and nuclei.

When we put radiation and materials together, a great variety of possible situations must be considered. Bombarding particles may have low or high energy; they may be charged, uncharged, or photons; they may be heavy or light in the scale of masses. The targets may be similarly distinguished, but also exhibit degrees of binding that range from none ("free" particles), to weak (atoms in molecules and electrons in atoms), to strong (nucleons in nuclei). In most interactions, the higher the projectile energy in comparison with the energy of binding of the structure, the greater is the effect.

Out of the broad subject we shall select for review some of the reactions that are important in the nuclear energy field. Looking ahead, we shall need to understand the effects produced by the particles and rays from radioactivity and other nuclear reactions. Materials affected may be in or around a nuclear reactor, as part of its construction or inserted to be irradiated. Materials may be of biological form, including the human body, or they may be inert substances used for protective shielding

53

against radiation. We shall not attempt to explain the processes rigorously, but be content with qualitative descriptions based on analogy with collisions viewed on an elementary physics level.

6.1 EXCITATION AND IONIZATION BY ELECTRONS

These processes occur in the familiar fluorescent lightbulb, or in a vacuum tube used in electrical devices, in an X-ray machine, or in matter exposed to beta particles. If an electron that enters a material has a very low energy, it will merely migrate without affecting the molecules significantly. If its energy is larger, it may impart energy to atomic electrons as described by the Bohr theory (Chapter 2), causing excitation of electrons to higher energy states or producing ionization, with subsequent emission of light. When electrons of inner orbits in heavy elements are displaced, the resultant high energy radiation is classed as X-rays. These rays, which are so useful for internal examination of the human body, are produced by accelerating electrons in a vacuum chamber to energies in the kilovolt range and allowing them to strike a heavy element target. In addition to the X-rays due to transitions in the electronic orbits, a similar radiation called *bremsstrahlung* (German: braking radiation) is produced. It arises from the deflection and resulting acceleration of electrons as they encounter nuclei or atomic electrons.

Beta particles as electrons from nuclear reactions have energies in the range 0.01–1 MeV, and thus are capable of producing large amounts of ionization as they penetrate a substance. As a rough rule of thumb, about 32 eV of energy is required to produce one ion pair. The beta particles lose energy with each event, and eventually are stopped. For electrons of 1 MeV energy, the range, as the typical distance of penetration, is no more than a few millimeters in liquids and solids or a few meters in air.

6.2 HEAVY CHARGED PARTICLE SLOWING BY ATOMS

Charged particles such as protons, alpha particles, or ions such as the fragments of fission are classed as heavy particles, being much more massive than the electron. For the same particle energy they have far less speed than an electron, but they are less readily deflected in their motion than electrons because of their inertia. The mechanism by which heavy ions slow down in matter is primarily electrostatic interaction with atomic electrons. As the positively charged projectile approaches and passes, with the attraction to electrons varying with distance of separation as

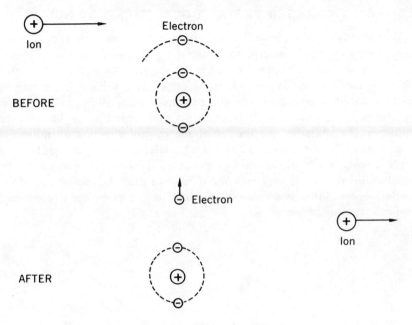

Fig. 6.1. Interaction of heavy ion with electron.

$1/r^2$, the electron is displaced and gains energy, while the heavy particle loses energy. Figure 6.1 shows conditions before and after the collision schematically. It is found that the kinetic energy lost in one collision is proportional to the square of Z, the number of external electrons in the target atom, and inversely proportional to the kinetic energy of the projectile. A great deal of ionization is produced by the heavy ion as it moves through matter. Although the projectile of energy in the million-electron-volt range loses only a small fraction of its energy in one collision, the amount of energy imparted to the electron can be large compared with its binding to the atom or molecule, and the electron is completely removed. As the result of these interactions, the energy of the heavy ion is reduced and it eventually is stopped in a range that is very much shorter than that for electrons. A 2 MeV alpha particle has a range of about 1 cm in air; it can be stopped by a sheet of paper or the outer layer of skin of the body. Because of these short ranges, there is little difficulty in providing protective shielding against alpha particles.

6.3 HEAVY CHARGED PARTICLE SCATTERING BY NUCLEI

When a high-speed charged ion such as an alpha particle encounters a very heavy charged nucleus, the mutual repulsion of the two particles causes the projectile to move on a hyperbolic path, as in Fig. 6.2. Such a collision can take place in an ionized gas or in a solid if the incoming particle avoids interaction with the external electrons of the atom. The projectile is scattered through an angle that depends on the detailed nature of the collision, i.e., the initial energy and direction of motion of the incoming ion relative to the target nucleus, and the magnitude of electric charge of the interacting particles. Unless the energy of the bombarding particle is very high and it comes within the short range of the nuclear force, there is a small chance that it can enter the nucleus and cause a nuclear reaction.

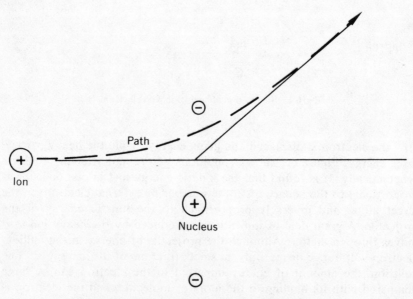

Fig. 6.2. Interaction of heavy ion with nucleus.

6.4 GAMMA RAY INTERACTIONS WITH MATTER

We now turn to a group of three related processes involving gamma ray photons produced by nuclear reactions. These have energies as high as a few MeV. The interactions include simple scattering of the photon, ionization by it, and a special nuclear reaction known as pair production.

(a) Photon–Electron Scattering

One of the easiest processes to visualize is the interaction of a photon of energy $E = h\nu$ and an electron of rest mass m_0. Although the electrons in a target atom can be regarded as moving and bound to their nucleus, the energies involved are very small (eV) in comparison with those of typical gamma rays (keV or MeV). Thus the electrons may be viewed as free stationary particles. The collision may be treated by the physical principles of energy and momentum conservation. As sketched in Fig. 6.3, the photon is deflected in its direction and loses energy, becoming a photon of new energy $E' = h\nu'$. The electron gains energy and moves away with high speed v and total mass-energy mc^2, leaving the atom ionized. In this *Compton effect*, named after its discoverer, one finds that the greatest photon energy loss occurs when it is scattered backward (180°) from the original direction. Then, if E is much larger than the rest energy of the electron $E_0 = m_0 c^2 = 0.51$ MeV, it is found that the *final photon energy E'* is equal to $E_0/2$. On the other hand, if E is much smaller than E_0, the *fractional energy loss* of the photon is $2E/E_0$ (see also Problem 6.3). The derivation of the photon energy loss in general is complicated by the fact that the special theory of relativity must be applied.

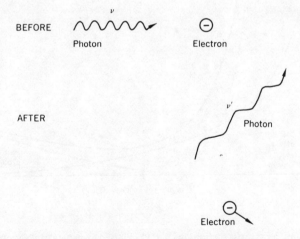

Fig. 6.3.   Photon–electron scattering (Compton effect).

The probability of Compton scattering is expressed by a cross section, which is smaller for larger gamma energies as shown in Fig. 6.4 for the element lead, a common material for shielding against X-rays or gamma rays. We can deduce that the chance of collision increases with each successive loss of energy by the photon, and eventually the photon disappears.

(b) Photoelectric Effect

This process is in competition with scattering. An incident photon of high enough energy dislodges an electron from the atom, leaving a positively charged ion. In so doing, the photon is absorbed and thus lost, see Fig. 6.5. The cross section for the photoelectric effect decreases with increasing photon energy, as sketched in Fig. 6.4 for the element lead.

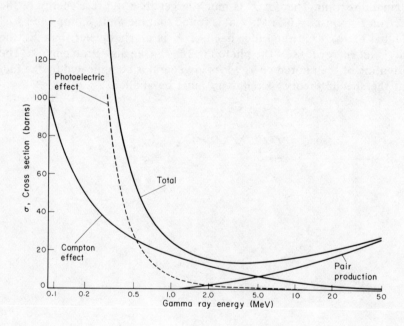

Fig. 6.4. Gamma ray cross sections in lead, Pb. Plotted from data in a National Bureau of Standards report NSRDS-NBS-29.

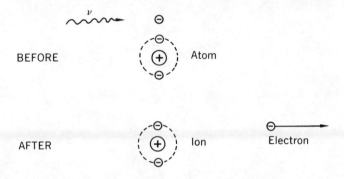

Fig. 6.5. Photoelectric effect.

The above two processes are usually treated separately even though both result in ionization. In the Compton effect, a photon of lower energy survives; but in the photoelectric effect, the photon is eliminated. In each case, the electron released may have enough energy to excite or ionize other atoms by the mechanism described earlier. Also, the ejection of the electron is followed by light emission or X-ray production, depending on whether an outer shell or inner shell is involved. The secondary particles and rays may be more important in the long run than the primary particle.

(c) Electron–Positron Pair Production

The third process to be considered is one in which the photon is converted into matter. This is entirely in accord with Einstein's theory of the equivalence of mass and energy. In the presence of a nucleus, as sketched in Fig. 6.6, a gamma ray photon disappears and two particles appear—an electron and a positron. Since these are of equal charge but of opposite sign, there is no net charge after the reaction, just as before, the gamma ray having zero charge. The law of conservation of charge is thus met. The total new mass produced is twice the mass-energy of the electron, $2(0.51) = 1.02 \text{ MeV}$, which means that the reaction can occur only if the gamma ray has at least this amount of energy. The cross section for the process of pair production rises from zero as shown in Fig. 6.4 for lead. The reverse process also takes place. As sketched in Fig. 6.7, when an electron and a positron combine, they are annihilated as material particles, and two gamma rays of energy totaling at least 1.02 MeV are released. That there must be two photons is a consequence of the principle of momentum conservation.

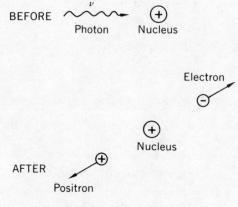

Fig. 6.6.   Pair production.

We note that the total gamma ray cross section curve for a substance, as the sum of the components for Compton effect, photoelectric effect, and pair production, exhibits a minimum around 3 MeV energy. This implies that gamma rays in this vicinity are more penetrating than those of higher or lower energy. In contrast with the case of beta particles and

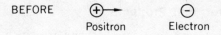

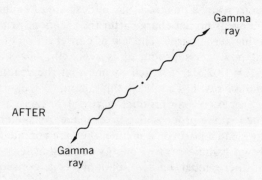

Fig. 6.7.   Pair annihilation.

alpha particles, which have a definite range, a certain fraction of incident gamma rays can pass through any thickness of material. The exponential expression $e^{-\Sigma z}$ as used to describe neutron behavior can be carried over to the attenuation of gamma rays in matter. One can use the mean free path $\lambda = 1/\Sigma$ or better, the half-thickness $0.693/\Sigma$, as the distance in which the intensity of a gamma ray beam is reduced by a factor of two.

6.5 NEUTRON REACTIONS

For completeness, we include a mention of the interaction of neutrons with matter. As discussed in Chapters 4 and 5, neutrons may be scattered by nuclei elastically or inelastically, may be captured with resulting gamma ray emission, or may cause fission. If this energy is high enough, neutrons may induce (n, p) and (n, α) reactions as well.

We are now in a position to understand the connection between neutron reactions and atomic processes. When a high-speed neutron strikes the hydrogen atom in a water molecule, a proton is ejected, resulting in chemical dissociation of the H_2O. A similar effect takes place in molecules of cells in any biological tissue. Now, the proton as a heavy charged particle passes through matter, slowing and creating ionization along its path. Thus two types of radiation damage take place—primary and secondary.

After many collisions, the neutron arrives at a low enough energy that it can be readily absorbed. If it is captured by the proton in a molecule of water or some other hydrocarbon, a gamma ray is released, as discussed in Chapter 4. The resulting deuteron recoils with energy that is much smaller than that of the gamma ray, but still is far greater than the energy of binding of atoms in the water molecule. Again dissociation of the compound takes place, which can be regarded as a form of radiation damage.

6.6 SUMMARY

Radiation of special interest includes electrons, heavy charged particles, photons, and neutrons. Each of the particles tends to lose energy by interaction with the electrons and nuclei of matter, and each creates ionization in different degrees. The ranges of beta particles and alpha particles are short, but gamma rays penetrate in accord with an exponential law. Gamma rays can also produce electron–positron pairs. Neutrons of both high and low energy can create radiation damage in molecular materials.

6.7 PROBLEMS

6.1. The charged particles in a highly ionized electrical discharge in hydrogen gas—protons and electrons, mass ratio $m_p/m_e = 1836$—have the same energies. What is the ratio of the speeds v_p/v_e? Of the momenta p_p/p_e?

6.2. A gamma ray from neutron capture has an energy of 6 MeV. What is its frequency? Its wavelength?

6.3. For 180° scattering of gamma rays or X-rays by electrons, the final energy of the photon is

$$E' = \frac{1}{\dfrac{1}{E} + \dfrac{2}{E_0}}.$$

(a) What is the final photon energy for the 6 MeV gamma ray of Problem 6.2?

(b) Verify that if $E \gg E_0$, then $E' \simeq E_0/2$ and if $E \ll E_0$, $(E - E')/E \simeq 2E/E_0$.

(c) Which approximation should be used for a 6 MeV gamma ray? Verify numerically.

6.4. An electron–positron pair is produced by a gamma ray of 2.26 MeV. What is the kinetic energy imparted to each of the charged particles?

6.5. Estimate the thickness of paper required to stop 2 MeV alpha particles, assuming the paper to be of density 1.29 g/cm^3, about the same electronic composition as air, density 1.29×10^{-3} g/cm^3.

6.6. The element lead, $M = 206$, has a density of 11.3 g/cm^3. Find the number of atoms per cubic centimeter. If the total gamma ray cross section at 3 MeV is 14 barns, what is the macroscopic cross section Σ and the half-thickness $0.693/\Sigma$?

6.7. The range of beta particles of energy greater than 0.8 MeV is given roughly by the relation

$$R(\text{cm}) = \frac{0.55E(\text{MeV}) - 0.16}{\rho(\text{g/cm}^3)}.$$

Find what thickness of aluminum sheet (density 2.7 g/cm^3) is enough to stop the betas from phosphorus-32 (see Table 3.1).

6.8. A radiation worker's hands are exposed for 5 seconds to a 3×10^8 cm^{-2} sec^{-1} beam of 1 MeV beta particles. Find the range in tissue of density 1.0 g/cm^3 and calculate the amounts of charge and energy deposition in C/cm^3 and J/g. Note that the charge on the electron is 1.60×10^{-19} C.

7

Fission

Out of the many nuclear reactions known, that resulting in fission has the greatest practical significance. In this chapter we shall describe the mechanism of the process, identify the byproducts, introduce the concept of the chain reaction, and look at the energy yield from the consumption of nuclear fuels.

7.1 THE FISSION PROCESS

The absorption of a neutron by most isotopes involves radiative capture, with the excitation energy appearing as a gamma ray. In certain heavy elements, notably uranium and plutonium, an alternate consequence is observed—the splitting of the nucleus into two massive fragments, a process called fission. Figure 7.1 shows the sequence of events, using the reaction with U-235 to illustrate. In Stage A, the neutron approaches the U-235 nucleus. In Stage B, the U-236 nucleus has been formed, in an excited state. The excess energy in some interactions may be released as a gamma ray, but more frequently, the energy causes distortions of the nucleus into a dumbbell shape, as in Stage C. The parts of the nucleus oscillate in a manner analogous to the motion of a drop of liquid. Because of the dominance of electrostatic repulsion over nuclear attraction, the two parts can separate, as in Stage D. They are then called fission fragments, bearing most of the mass-energy released. They fly apart at high speeds, carrying some 166 MeV of kinetic energy out of the total of around 200 MeV released in the whole process. As the fragments separate, they lose atomic electrons, and the resulting high-speed ions lose

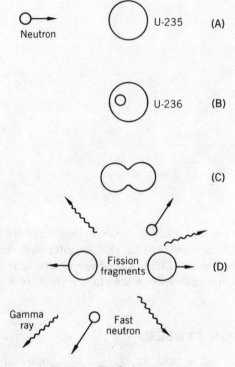

Fig. 7.1. The fission process.

energy by interaction with the atoms and molecules of the surrounding medium. The resultant thermal energy is recoverable if the fission takes place in a nuclear reactor. Also shown in the diagram are the gamma rays and fast neutrons that come off at the time of splitting.

7.2 ENERGY CONSIDERATIONS

The absorption of a neutron by a nucleus such as U-235 gives rise to extra internal energy of the product, because the sum of masses of the two interacting particles is greater than that of a normal U-236 nucleus. We write the first step in the reaction

$$^{235}_{92}\text{U} + ^{1}_{0}\text{n} \rightarrow (^{236}_{92}\text{U})^*,$$

where the asterisk signifies the excited state. The mass in atomic mass units of (U-236)* is the sum $235.043925 + 1.008665 = 236.052590$. However,

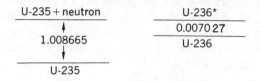

Fig. 7.2. Excitation energy due to neutron absorption.

U-236 in its ground state has a mass of only 236.045563, lower by 0.007027 amu or 6.5 MeV. This amount of excess energy is sufficient to cause fission. Figure 7.2 shows these energy relations.

The above calculation did not include any kinetic energy brought to the reaction by the neutron, on the grounds that fission can be induced by absorption in U-235 by very slow neutrons. Only one natural isotope, $^{235}_{92}U$, undergoes fission in this way, while $^{239}_{94}Pu$ and $^{233}_{92}U$ are the main artificial isotopes that do so. Most other heavy isotopes require significantly larger excitation energy to bring the compound nucleus to the required energy level for fission to occur, and the extra energy must be provided by the motion of the incoming neutron. For example, neutrons of at least 0.9 MeV are required to cause fission from U-238, and other isotopes require even higher energy. The precise terminology is as follows: *fissile* materials are those giving rise to fission with slow neutrons; many isotopes are *fissionable*, if enough energy is supplied. It is advantageous to use fast neutrons—of the order of 1 MeV energy—to cause fission. As will be discussed in Chapter 15, the fast reactor permits the "breeding" of nuclear fuel. In a few elements such as californium, spontaneous fission takes place. The isotope $^{252}_{98}Cf$, produced artificially by a sequence of neutron absorption, has a half-life of 2.646 yr, decaying by alpha emission (97%) and spontaneous fission (3%).

It is not obvious how the introduction of only 6.5 MeV of excitation energy can produce a reaction yielding as much as 200 MeV. The excitation triggers the separation of the two fragments (plus neutrons) with total mass much less than the mass of the nucleus from which they come (see Problem 7.4).

7.3 BYPRODUCTS OF FISSION

Accompanying the fission process is the release of several neutrons, which are all-important for the practical application to a self-sustaining chain reaction. The number that appear ν (nu) ranges from 1 to 7, with an average in the range 2 to 3 depending on the isotope and the bombarding neutron energy. For example, in U-235 with slow neutrons the average number $\bar{\nu}$ is 2.43. Most of these are released instantly, the so-called *prompt neutrons*, while a small percentage, 0.65% for U-235, appear later as the result of radioactive decay of certain fission fragments. These *delayed neutrons* provide considerable inherent safety and controllability in the operation of nuclear reactors, as we shall see later.

The nuclear reaction equation for fission resulting from neutron absorption in U-235 may be written in general form, letting the chemical symbols for the two fragments be labeled F_1 and F_2 to indicate many possible ways of splitting. Thus

$$^{235}_{92}U + ^{1}_{0}n \rightarrow ^{A_1}_{Z_1}F_1 + ^{A_2}_{Z_2}F_2 + \nu^{1}_{0}n + \text{energy.}$$

The appropriate mass numbers and atomic numbers are attached. One example, in which the fission fragments are isotopes of krypton and barium, is

$$^{235}_{92}U + ^{1}_{0}n \rightarrow ^{90}_{36}Kr + ^{144}_{56}Ba + 2\,^{1}_{0}n + E.$$

Mass numbers ranging from 75 to 160 are observed, with the most probable at around 92 and 144 as sketched in Fig. 7.3. The ordinate on this graph is the percentage yield of each mass number, e.g., about 6% for mass numbers 90 and 144. If the number of fissions is given, the number of atoms of those types are 0.06 as large.

As a collection of isotopes, these byproducts are called fission products. The isotopes have an excess of neutrons or a deficiency of protons in comparison with naturally occurring elements. For example, the main isotope of barium is $^{137}_{56}Ba$, and a prominent element of mass 144 is $^{144}_{60}Nd$. Thus there are 7 extra neutrons or 4 too few protons in the barium isotope from fission, and it is highly unstable. Radioactive decay, usually involving several emissions of beta particles and delayed gamma

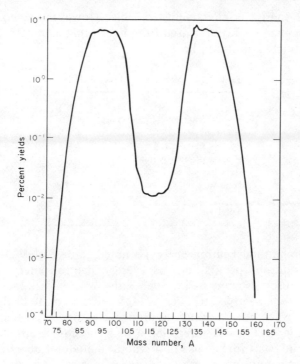

Fig. 7.3.   Yield of fission products according to mass number. (Courtesy of T. R. England of Los Alamos Scientific Laboratory)

rays in a chain of events, brings the particles down to stable forms. An example is

$$\ce{^{90}_{36}Kr ->[33\,sec] ^{90}_{37}Rb ->[2.91\,min] ^{90}_{38}Sr ->[27.7\,yr] ^{90}_{39}Y ->[64\,hr] ^{90}_{40}Zr.}$$

The hazard associated with the radioactive emanations from fission products is evident when we consider the large yields and the short half-lives.

The total energy release from fission, after all of the particles from decay have been released, is about 200 MeV. This is distributed among the various processes as shown in Table 7.1. The "prompt" gamma rays are emitted as a part of fission, the rest are "decay" gammas. Neutrinos accompany the beta particle emission, but since they are such highly penetrating particles their energy cannot be counted as part of the useful thermal energy yield of the fission process. Thus only about 190 MeV of the fission energy is effectively available. However, several MeV of energy from gamma rays released from nuclei that capture neutrons can also be extracted as useful heat.

Table 7.1.   Energy from Fission, U-235.

	MeV
Fission fragment kinetic energy	166
Neutrons	5
Prompt gamma rays	7
Fission product gamma rays	7
Beta particles	5
Neutrinos	10
Total	200

The average total neutron energy is noted to be 5 MeV. If there are about 2.5 neutrons per fission, the average neutron energy is 2 MeV. When one observes many fission events, the neutrons are found to range in energy from nearly 0 to over 10 MeV, with a most likely value of 0.7 MeV. We note that the neutrons produced by fission are fast, while the cross section for the fission reaction is high for slow neutrons. This fact serves as the basis for the use of a reactor moderator containing a light element that permits neutrons to slow down, by a succession of collisions, to an energy favorable for fission.

Although fission is the dominant process, a certain fraction of the absorptions of neutrons in uranium merely result in radiative capture, according to

$$^{235}_{92}U + {}^{1}_{0}n \rightarrow {}^{236}_{92}U + \gamma.$$

The U-236 is relatively stable, having a half-life of 2.34×10^7 yr. About 14% of the absorptions are of this type, with fission occurring in the remaining 86%. This means that η (eta), the number of neutrons produced per *absorption* in U-235 is lower than ν, the number per *fission*. Thus using $\bar{\nu} = 2.42$, η is $(0.86)(2.42) = 2.07$. The effectiveness of any nuclear fuel is sensitively dependent on the value of η. We find that η is larger for fission induced by fast neutrons than that by slow neutrons.

The possibility of a chain reaction was recognized as soon as it was known that neutrons were released in the fission process. If a neutron is absorbed by the nucleus of one atom of uranium and one neutron is produced, the latter can be absorbed in a second uranium atom, and so on. In order to sustain a chain reaction as in a nuclear reactor or in a nuclear weapon, the value of η must be somewhat above one because of processes that compete with absorption in uranium, such as capture in other materials and escape from the system. The size of η has two important consequences. First, there is a possibility of a growth of neutron population with time. After all extraneous absorption and losses have been accounted for, if one absorption in uranium ultimately gives rise to say 1.1 neutrons, these can be absorbed to give $(1.1)(1.1) = 1.21$, which produce 1.331, etc. The number available increases rapidly with time. Second, there is a possibility of using the extra neutron, over and above the one required to maintain the chain reaction, to produce new fissile materials. "Conversion" involves the production of some new nuclear fuel to replace that used up, while "breeding" is achieved if more fuel is produced than is used.

Out of the hundreds of isotopes found in nature, only one is fissile, $^{235}_{92}\text{U}$. Unfortunately, it is the less abundant of the isotopes of uranium, with weight percentage in natural uranium of only 0.711, in comparison with 99.3% of the heavier isotope $^{238}_{92}\text{U}$. The two other most important fissile materials, plutonium-239 and uranium-233, are "artificial" in the sense that they are man-made by use of neutron irradiation of two *fertile* materials, respectively, uranium-238 and thorium-232. The reactions by which $^{239}_{94}\text{Pu}$ is produced are

$$^{238}_{92}\text{U} + {}^{1}_{0}\text{n} \rightarrow {}^{239}_{92}\text{U},$$

$$^{239}_{92}\text{U} \xrightarrow[23.5 \text{ min}]{} {}^{239}_{93}\text{Np} + {}^{0}_{-1}\text{e},$$

$$^{239}_{93}\text{Np} \xrightarrow[2.35 \text{ day}]{} {}^{239}_{94}\text{Pu} + {}^{0}_{-1}\text{e},$$

while those yielding $^{233}_{92}\text{U}$ are

$$^{232}_{90}\text{Th} + {}^{1}_{0}\text{n} \rightarrow {}^{233}_{90}\text{Th},$$

$$^{233}_{90}\text{Th} \xrightarrow[22.3 \text{ min}]{} {}^{233}_{91}\text{Pa} + {}^{0}_{-1}\text{e},$$

$$^{233}_{91}\text{Pa} \xrightarrow[27 \text{ day}]{} {}^{233}_{92}\text{U} + {}^{0}_{-1}\text{e}.$$

The half-lives for decay of the intermediate isotopes are short compared with times involved in the production of these fissile materials; and for many purposes, these decay steps can be ignored. It is important to note that although uranium-238 is not fissile, it can be put to good use as a fertile material for the production of plutonium-239, so long as there are enough free neutrons available.

7.4 ENERGY FROM NUCLEAR FUELS

The practical significance of the fission process is revealed by calculation of the amount of uranium that is consumed to obtain a given amount of energy. Each fission yields 190 MeV of useful energy, which is also $(190 \text{ MeV})(1.60 \times 10^{-13} \text{ J/MeV}) = 3.04 \times 10^{-11} \text{J}$. Thus the number of fissions required to obtain 1 W-sec of energy is $1/(3.04 \times 10^{-11}) = 3.3 \times 10^{10}$. The number of U-235 atoms consumed in a thermal reactor is larger by the factor $1/0.86 = 1.16$ because of the formation of U-236 in part of the reactions.

In one day's operation of a reactor per megawatt of thermal power, the number of U-235 nuclei burned is

$$\frac{(10^6 \text{ W})(3.3 \times 10^{10} \text{ fissions/W-sec})(86,400 \text{ sec/day})}{0.80 \text{ fissions/absorptions}}$$

$$= 3.32 \times 10^{21} \text{ absorptions/day.}$$

Then since 235 g corresponds to Avogadro's number of atoms 6.02×10^{23}, the U-235 weight consumed at 1 MW power is

$$\frac{(3.32 \times 10^{21} \text{ day}^{-1})(235 \text{ g})}{(6.02 \times 10^{23})} \simeq 1.3 \text{ g/day.}$$

In other words, 1.3 g of fuel is used per megawatt-day of useful thermal energy released. In a typical reactor, which produces 3000 MW of thermal power, the U-235 fuel consumption is about 4 kg/day. To produce the same energy by use of fossil fuels such as coal, oil, or gas, millions of times as much weight would be required.

7.5 SUMMARY

Neutron absorption by the nuclei of heavy elements gives rise to fission, in which heavy fragments, fast neutrons, and other radiations are released. Fissile materials are natural U-235 and man-made isotopes Pu-239 and U-233. Many different radioactive isotopes are released in the fission process, and more neutrons are produced than are used, which makes possible a chain reaction and under certain conditions "conversion" and "breeding" of new fuels. Useful energy amounts to 190 MeV per fission, requiring only 1.3 g of U-235 to be consumed to obtain 1 MW-day of energy.

7.6 PROBLEMS

7.1. Calculate the excitation energy in (Pu-240)* resulting from the absorption of a neutron of mass 1.008665 in Pu-239, mass 239.052158 assuming the mass of Pu-240 to be 240.053809.

7.2. If three neutrons and a xenon-133 atom ($^{133}_{54}$Xe) are produced when a U-235 atom is bombarded by a neutron, what is the second fission product isotope?

7.3. Neglecting neutron energy and momentum effects, what are the energies of the two fission fragments if their mass ratio is $3:2$?

7.4. Calculate the energy yield from the reaction

$$^{235}_{92}U + {}^{1}_{0}n \rightarrow {}^{140}_{55}Cs + {}^{92}_{37}Rb + 4\,{}^{1}_{0}n + E$$

using atomic masses 139.91709 for cesium and 91.91935 for rubidium.

7.5. The value of η for U-233 for thermal neutrons is approximately 2.29. Using the cross sections for capture $\sigma_c = 46$ barns and fission $\sigma_f = 530$ barns, deduce the value of ν, the number of neutrons per fission.

7.6. A mass of 8000 kg of slightly enriched uranium (2% U-235, 98% U-238) is exposed for 30 days in a reactor operating at heat power 2000 MW. Neglecting consumption of U-238, what is the final fuel composition?

7.7. The per capita consumption of electrical energy in the United States is about 50 kWh/day. If this were provided by fission with $\frac{2}{3}$ of the heat wasted, how much U-235 would each person use per day?

7.8. Calculate the number of kilograms of coal, oil, and natural gas that must be burned each day to operate a 3000-MW thermal power plant, which consumes 4 kg/day of uranium-235. The heats of combustion of the three fuels (in kJ/g) are, respectively, 32, 44, and 50.

8

Fusion

When two light nuclear particles combine or "fuse" together, energy is released because the product nuclei have less mass than the original particles. Such fusion reactions can be caused by bombarding targets with charged particles, using an accelerator, or by raising the temperature of a gas to a high enough level for nuclear reactions to take place. In this chapter we shall describe the interactions in the microscopic sense and discuss the phenomena that affect our ability to achieve a practical large-scale source of energy from fusion.

8.1 FUSION REACTIONS

The possibility of release of large amounts of nuclear energy can be seen by comparing the masses of nuclei of low atomic number. Suppose that one could combine two hydrogen nuclei and two neutrons to form the helium nucleus. In the reaction

$$2\,{}^{1}_{1}\text{H} + 2\,{}^{1}_{0}\text{n} \rightarrow {}^{4}_{2}\text{He},$$

the mass-energy difference (using atom masses) is

$$2(1.007825) + 2(1.008665) - 4.002603 = 0.030377 \text{ amu},$$

which corresponds to 28.2 MeV energy. A comparable amount of energy would be obtained by combining four hydrogen nuclei to form helium plus two positrons

$$4\,{}^{1}_{1}\text{H} \rightarrow {}^{4}_{2}\text{He} + 2\,{}^{0}_{1}\text{e}.$$

This reaction in effect takes place in the sun and other stars through the so-called carbon cycle, a complicated chain of events involving hydrogen and isotopes of the elements carbon, oxygen, and nitrogen. The cycle is extremely slow, however, and is not suitable for terrestrial application. The sun's large energy yield is due to its tremendous mass of materials, not to the large rate of nuclear reactions per unit volume.

In the "hydrogen bomb," on the other hand, the high temperatures created by a fission reaction cause the fusion reaction to proceed in a rapid and uncontrolled manner. Between these extremes is the possibility of achieving a controlled fusion reaction that utilizes inexpensive and abundant fuels. As yet, a practical fusion device has not been developed, and considerable research and development will be required to reach that goal. Let us now examine the nuclear reactions that might be employed. There appears to be no mechanism by which four separate nuclei can be made to fuse directly, and thus combinations of two particles must be sought.

The most promising reactions make use of the isotope deuterium. It is present in hydrogen as in water with abundance only 0.015%, i.e., there is one atom of ^2_1H for every 6700 atoms of ^1_1H, but since our planet has enormous amounts of water, the fuel available is almost inexhaustible. Four reactions are important:

$$^2_1H + {}^2_1H \rightarrow {}^3_1H + {}^1_1H + 4.03 \text{ MeV},$$

$$^2_1H + {}^2_1H \rightarrow {}^3_2He + {}^1_0n + 3.27 \text{ MeV},$$

$$^2_1H + {}^3_1H \rightarrow {}^4_2He + {}^1_0n + 17.6 \text{ MeV},$$

$$^2_1H + {}^3_2He \rightarrow {}^4_2He + {}^1_1H + 18.3 \text{ MeV}.$$

The fusion of two deuterons—deuterium nuclei—in what is designated the D–D reaction results in two processes of equal likelihood. The other reactions yield more energy but involve the artificial isotopes tritium, ^3_1H, and helium-3. We note that the products of the first and second equations appear as reactants in the third and fourth equations. This suggests that a composite process might be feasible. Suppose that each of the reactions could be made to proceed at the same rate, along with twice the reaction of neutron capture in hydrogen

$$^1_1H + {}^1_0n \rightarrow {}^2_1H + 2.2 \text{ MeV}.$$

Adding all of the equations, we find that the net effect is to convert deuterium into helium according to

$$4\,^2_1H \rightarrow 2\,^4_2He + 47.7\,MeV.$$

The energy yield per atomic mass unit of deuterium fuel would thus be about 6 MeV, which is much more favorable than the yield per atomic mass unit of U-235 burned, which is only $190/235 = 0.85$ MeV.

8.2   ELECTROSTATIC AND NUCLEAR FORCES

The reactions described above do not take place merely by mixing the ingredients because of the very strong force of electrostatic repulsion between the charged nuclei. Only by giving one or both of the particles a high speed can they be brought close enough to each other for the strong nuclear force to dominate the electrical force. This behavior is in sharp contrast with the ease with which neutrons interact with nuclei.

There are two consequences of the fact that the coulomb force between two charges of atomic numbers Z_1 and Z_2 varies with separation R according to Z_1Z_2/R^2. First, we see that fusion is unlikely in elements other than those low in the periodic table. Second, the force and corresponding potential energy of repulsion is very large at the 10^{-15} m range of nuclear forces, and thus the chance of reaction is negligible unless particle energies are of the order of keV. Figure 8.1 shows the cross section for the D–D reaction. The strong dependence on energy is noted, with σ_{DD} rising by a factor of 1000 in the range 10–75 keV.

Energies in the kilo-electron-volt and million-electron-volt range can be achieved by a variety of charged particle accelerators. Bombardment of a solid or gaseous deuterium target by high-speed deuterons gives fusion reactions, but most of the particle energy goes into electrostatic interactions that merely heat up the bulk of the target. The amount of energy required to operate the accelerator greatly exceeds the recoverable fusion energy, and thus some other technique is required.

8.3   THERMONUCLEAR REACTIONS IN IONIC PLASMA

The most promising medium in which to obtain the high particle energies that are needed for practical fusion is the plasma. It consists of a completely ionized gas as in an electrical discharge created by the acceleration of electrons. Equal numbers of electrons and deuterons are present, making the medium electrically neutral. Through the injection of enough energy into the plasma its temperature can be increased, and the deuterons reach the speed for fusion to be favorable. The term

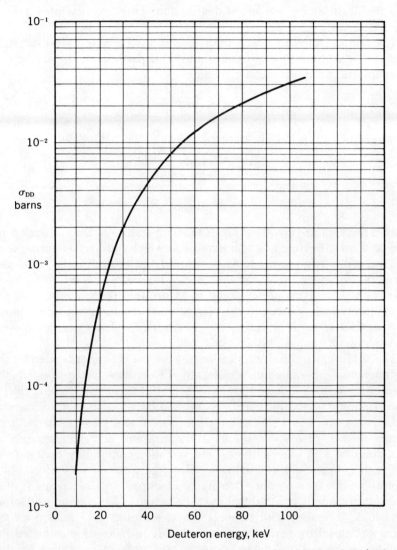

Fig. 8.1. Cross section for D–D reaction. (From Raymond L. Murray, *Introduction to Nuclear Engineering*, 2nd Ed. © 1961. Reprinted by permission of Prentice-Hall, Inc., Englewood Cliffs, New Jersey.)

thermonuclear is applied to reactions induced by high thermal energy, and the particles obey a speed distribution similar to that of a gas, as discussed in Chapter 2.

The temperatures to which the plasma must be raised are extremely high, as we can see by expressing an average particle energy in terms of temperature, using the kinetic relation

$$\bar{E} = \frac{3}{2} kT.$$

For example, even if $\bar{E}$ is as low as 10 keV, the temperature is

$$T = \frac{2}{3} \frac{(10^4 \text{ eV})(1.60 \times 10^{-19} \text{ J/eV})}{1.38 \times 10^{-23} \text{ J/}^\circ\text{K}}$$

or

$$T = 77,000,000^\circ\text{K}.$$

Such a temperature greatly exceeds the temperature of the surface of the sun, and is far beyond any temperature at which ordinary materials melt and vaporize. The plasma must be created and heated to the necessary temperature under the constraint of electric and magnetic fields. Such forces on the plasma are required to assure that thermal energy is not prematurely lost. Moreover, the plasma must remain intact long enough for many nuclear reactions to occur, which is difficult because of inherent instabilities of such highly charged media.

The achievement of a practical energy source is further limited by the phenomenon of radiation losses. In Chapter 6 we discussed the bremsstrahlung radiation produced when electrons experience acceleration. Conditions are ideal for the generation of such electromagnetic radiation since the high-speed electrons in the plasma at elevated temperature experience continuous accelerations and decelerations as they interact with other charges. The radiation can readily escape from the region, because the number of target particles is very small. In a typical plasma, the number density of electrons and deuterons is 10^{15}, which corresponds to a rarefied gas. The amount of radiation production (and loss) increases with temperature at a slower rate than does the energy released by fusion, as shown in Fig. 8.2. At what is called the "ignition temperature," the lines cross. Only for temperatures above that value, $400,000,000^\circ\text{K}$ in the case of the D–D reaction, will there be a net energy yield, assuming that the radiation is lost. In a later chapter we shall describe some of the devices that have been used to explore the possibility of achieving a fusion reactor.

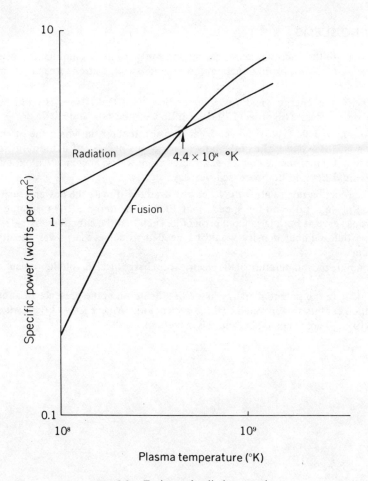

Fig. 8.2.   Fusion and radiation energies.

8.4 SUMMARY

Nuclear energy is released when nuclei of two light elements combine. The most favorable fusion reactions involve deuterium, which is a natural component of water and thus is a very abundant fuel. The reaction takes place only when the particles have a high enough speed to overcome the electrostatic repulsion of their charges. In a highly ionized electrical medium, the plasma, at temperatures of the order of 400,000,000°K, the fusion energy can exceed the energy loss due to radiation.

8.5 PROBLEMS

8.1. Calculate the energy release in amu and MeV from the combination of four hydrogen atoms to form a helium atom and two positrons (each of mass 0.000549 amu).

8.2. Verify the energy yield for the reaction $^2_1H + ^3_2He \rightarrow ^4_2He + ^1_1H + 18.3$ MeV, noting atomic masses (in order) 2.014102, 3.016029, 4.002603, and 1.007825.

8.3. To obtain 3000 MW of power from a fusion reactor, in which the effective reaction is $2\,^2_1H \rightarrow ^4_2He + 23.8$ MeV, how many grams per day of deuterium would be needed? If all of the ^2_1H could be extracted from water, how many kilograms of water would have to be processed per day?

8.4. The reaction rate relation $nvN\sigma$ can be used to estimate the power density of a fusion plasma. (a) Find the speed v_D of 100 keV deuterons. (b) Assuming that deuterons serve as both target and projectile, such that the effective v is $v_D/2$, find what particle number density would be needed to achieve a power density of 1 kW/cm^3.

8.5. Estimate the temperature of the electrical discharge in a 120-volt fluorescent light bulb.

8.6. Calculate the potential energy in eV of a deuteron in the presence of another when their centers are separated by three nuclear radii (*Note:* $E_p = kQ_1Q_2/R$ where $k = 9 \times 10^9$, Q's are in coulombs, and R is in meters).

Part II Nuclear Systems

The atomic and nuclear concepts we have described provide the basis for the operation of a number of devices, machines, or processes, ranging from very small radiation detectors to mammoth plants to process uranium or to generate electrical power. These systems may be designed to produce nuclear energy, or to make practical use of it, or to apply byproducts of nuclear reactions for beneficial purposes. In the next several chapters we shall explain the construction and operating principles of nuclear systems, referring back to basic concepts and looking forward to appreciating their impact on human affairs.

9

Particle Accelerators

A device that provides forces on charged particles by some combination of electric and magnetic fields and brings the ions to high speed and kinetic energy is called an accelerator. Many types have been developed for the study of nuclear reactions and basic nuclear structure, with an ever-increasing demand for higher particle energy. In this chapter we shall review the nature of the forces on charges and describe the arrangement and principle of operation of several important kinds of particle accelerators.

9.1 ELECTRIC AND MAGNETIC FORCES

Let us recall how charged particles are influenced by electric and magnetic fields. First, visualize a pair of parallel metal plates separated by a distance d as in the sample capacitor shown in Fig. 9.1. A potential difference V and electric field $\mathscr{E} = V/d$ are provided to the region of low gas pressure by a direct-current voltage supply such as a battery. If an electron of mass m and charge e is released at the negative plate, it will experience a force $\mathscr{E}e$, and its acceleration will be $\mathscr{E}e/m$. It will gain speed, and on reaching the positive plate it will have reached a kinetic energy $\frac{1}{2}mv^2 = Ve$. Thus its speed is $v = \sqrt{2Ve/m}$. For example, if V is 100 volts, the speed of an electron ($m = 9.1 \times 10^{-31}$ kg and $e = 1.60 \times 10^{-19}$ coulombs) is easily found to be 5.9×10^6 m/sec.

Next, let us introduce a charged particle of mass m, charge e, and speed v into a region with uniform magnetic field B, as in Fig. 9.2. If the charge enters in the direction of the field lines, it will not be affected, but if it

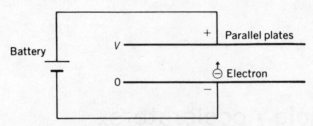

Fig. 9.1. Capacitor as accelerator.

enters perpendicularly to the field, it will move at constant speed on a circle. Its radius, called the radius of gyration, is $r = mv/eB$, such that the stronger the field or the lower the speed, the smaller will be the radius of motion. Letting the angular speed be $w = v/r$ we see that $w = eB/m$. If the charge enters at some other angle, it will move in a path called a helix, like a wire door spring.

Finally, let us release a charge in a region where the magnetic field B is changing with time. If the electron were inside the metal of a circular loop of wire of area A as in Fig. 9.3, it would experience an electric force induced by the change in magnetic flux BA. The same effect would take place without the presence of the wire, of course.

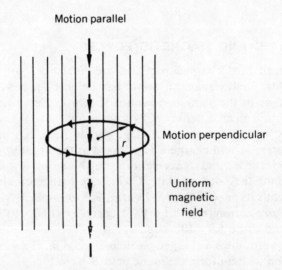

Fig. 9.2.   Electric charge motion in uniform magnetic field.

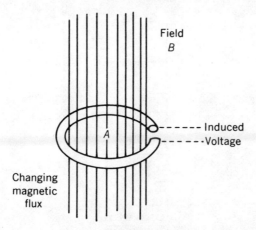

Fig. 9.3.   Magnetic induction.

9.2   HIGH-VOLTAGE MACHINES

One way to accelerate ions to high speed is to provide a large potential difference between a source of charges and a target. In effect, the phenomenon of lightning, in which a discharge from charged clouds to the earth takes place, is produced in the laboratory. Two devices of this type are commonly used. The first is the voltage multiplier or Cockroft–Walton machine, Fig. 9.4, which has a circuit that charges capacitors in parallel and discharges them in series.

The second is the electrostatic generator or Van de Graaff accelerator, the principle of which is sketched in Fig. 9.5. An insulated metal shell is raised to high potential by bringing it charge on a moving belt, permitting the acceleration of positive charges such as protons or deuterons. Particle energies of the order of 5 MeV are possible, with a very small spread in energy.

9.3   LINEAR ACCELERATOR

Rather than giving a charge one large acceleration with a high voltage, it can be brought to high speed by a succession of accelerations through relatively small potential differences, as in the linear accelerator, sketched in Fig. 9.6. It consists of a series of accelerating electrodes in the form of tubes with alternating electric potentials applied as shown. An electron or ion gains energy in the gaps between tubes and "drifts" without change of

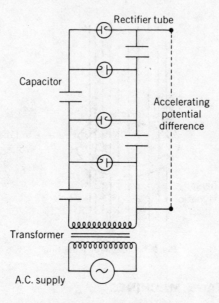

Fig. 9.4. Cockroft–Walton circuit.

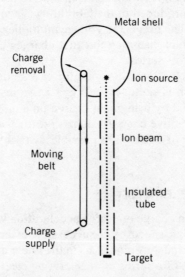

Fig. 9.5. Van de Graaff accelerator.

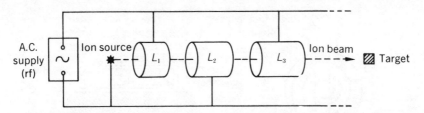

Fig. 9.6. Simple linear accelerator. (From Raymond L. Murray and Grover C. Cobb, *PHYSICS: Concepts and Consequences,* © 1970. Reprinted by permission of Prentice-Hall, Inc., Englewood Cliffs, New Jersey.)

energy while inside the tube, where the field is nearly zero. By the time the charge reaches the next gap, the voltage is again correct for acceleration. Because the ion is gaining speed along the path down the row of tubes, their lengths l must be successively longer in order for the time of flight in each to be constant. The time to go a distance l is l/v, which is equal to the half-period of the voltage cycle $T/2$. Particle energies of 20 billion electron-volts are obtained in the two-mile-long Stanford accelerator.

9.4 CYCLOTRON

Successive electrical accelerations by electrodes and circular motion within a magnetic field are combined in the cyclotron. As sketched in Fig. 9.7, ions such as protons, deuterons, or alpha particles are provided by a source at the center of a vacuum chamber located between the poles of a large electromagnet. Two hollow metal boxes called "dees" (in the shape of the letter D) are supplied with alternating voltages in correct frequency and opposite polarity. In the gap between dees, an ion gains energy as in the linear accelerator, then moves on a circle while inside the field-free region, guided by the magnetic field. Each crossing of the gap with potential difference V gives impetus to the ion with an energy gain Ve, and the radius of motion increases according to $r = v/w$, where $\omega = eB/m$ is the angular speed. The unique feature of the cyclotron is that the time required for one complete revolution, $T = 2\pi/\omega$, is independent of the radius of motion of the ion. Thus it is possible to use a synchronized alternating potential of constant frequency ν, angular frequency $\omega = 2\pi\nu$, to provide acceleration at the right instant.

For example, in a magnetic field B of 0.5 Wb/m^2 the angular speed for

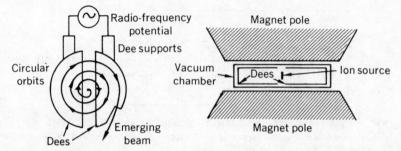

Fig. 9.7. Cyclotron. (From Raymond L. Murray and Grover C. Cobb, *PHYSICS: Concepts and Consequences*, © 1970. Reprinted by permission of Prentice-Hall, Inc., Englewood Cliffs, New Jersey.)

deuterons of mass 3.3×10^{-27} kg and charge 1.6×10^{-19} coulombs is

$$\omega = \frac{eB}{m} = \frac{(1.6 \times 10^{-19})(0.5)}{3.3 \times 10^{-27}} = 2.4 \times 10^{7}/\text{sec}.$$

Equating this to the angular frequency for the power supply, $\omega = 2\pi\nu$, we find $\nu = (2.4 \times 10^{7})/2\pi = 3.8 \times 10^{6}$ sec^{-1}, which is in the radio-frequency range.

The path of ions is approximately a spiral. When the outermost radius is reached and the ions have full energy, a beam is extracted from the dees by special electric and magnetic fields, and allowed to strike a target, in which nuclear reactions take place.

9.5 BETATRON

Electrons are brought to high speeds in the induction accelerator or betatron. A changing magnetic flux provides an electric field and a force on the charges, while they are guided in a path of constant radius. Figure 9.8 shows the vacuum chamber in the form of a doughnut placed between specially shaped magnetic poles. The force on electrons of charge e is in the direction tangential to the orbit of radius r. The rate at which the average magnetic field within the loop changes is $\Delta B/\Delta t$, provided by varying the current in the coils of the electromagnet. The magnitude of the force is†

†To show this, note that the area within the circular path is $A = \pi r^2$ and the magnetic flux is $\Phi = BA$. According to Faraday's law of induction, if the flux changes by $\Delta\Phi$ in a time Δt, a potential difference around a circuit of $V = \Delta\Phi/\Delta t$ is produced. The corresponding electric field is $\mathscr{E} = V/2\pi r$, and the force is $e\mathscr{E}$. Combining, the relation quoted is obtained.

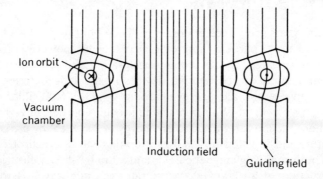

Fig. 9.8. Betatron. (From Raymond L. Murray and Grover C. Cobb, *PHYSICS: Concepts and Consequences*, © 1970. Reprinted by permission of Prentice-Hall, Inc., Englewood Cliffs, New Jersey.)

$$F = \frac{er}{2}\frac{\Delta B}{\Delta t}.$$

The charge continues to gain energy while remaining at the same radius if the magnetic field there is half the average field within the loop. The acceleration to energies in the million-electron-volt range takes place in the fraction of a second that it takes for the alternating magnetic current to go through a quarter-cycle.

The speeds reached in a betatron are high enough to require the use of relativistic formulas (Chapter 1). Let us find the mass m and speed v for an electron of kinetic energy $E_k = 1\,\mathrm{MeV}$. Rearranging the equation for kinetic energy, the ratio of m to the rest mass m_0 is

$$\frac{m}{m_0} = 1 + \frac{E_k}{m_0 c^2}.$$

Recalling that the rest energy $E_0 = mc^2$ for an electron is 0.51 MeV we obtain $m/m_0 = 1 + 1/0.51 = 2.96$. Solving Einstein's equation $m/m_0 = 1/\sqrt{1-(v/c)^2}$ for the speed, we find that $v = c\sqrt{1-(m_0/m)^2} = 0.94c$. Thus the 1 MeV electron's speed is close to that of light, $c = 3.0 \times 10^8\,\mathrm{m/sec}$, i.e., $v = 2.8 \times 10^8\,\mathrm{m/sec}$. If instead we impart a kinetic energy of 100 MeV to an electron, its mass increases by a factor 297 and its speed becomes $0.999995c$.

9.6 NEW DEVELOPMENTS

Over the years, new accelerators have been built to produce larger ion currents and higher particle energies. One of the most powerful accelerators in the world is located at the National Accelerator Laboratory near Chicago. By a combination of accelerating devices of the type described in this chapter, it gives protons an energy of around 400 GeV (400 billion electron-volts). First a Cockroft–Walton machine provides particle energies of 0.75 MeV. Then, the ions are raised to an energy of 200 MeV by use of a linear accelerator, and injected for final acceleration into a synchrotron, which involves both changing magnetic fields and radio-frequency electric fields.

Construction of an 800-GeV accelerator is underway at Brookhaven National Laboratory. Two 400-GeV proton accelerators are arranged concentrically such that beams of particles collide head on.

By the use of accelerators of greater sophistication and higher particle energy many new subnuclear particles such as mesons and xi, sigma, and lambda particles have been discovered and the internal structure of nuclei has become better understood.

It is possible that accelerators can be used directly to help solve the energy problem. Experiments at California radiation laboratories showed that large neutron yields were achieved in targets bombarded by charged particles such as deuterons or protons of several hundred MeV energy. New dramatic nuclear reactions are involved. One is the "stripping" reaction, Fig. 9.9(a), in which a deuteron is broken into a proton and a neutron by the impact on a target nucleus. Another is the process of "spallation" in which a nucleus is broken into pieces by an energetic projectile. Fig. 9.9(b) shows how a cascade of nucleons is produced by spallation. A third is "evaporation" in which neutrons fly out of a nucleus with some 100 MeV of internal excitation energy, see Fig. 9.9(c). The average energy of evaporation of neutrons is about 3 MeV. The excited nucleus may undergo fission, which releases neutrons, and further evaporation from the fission fragments can occur.

It has been predicted that as many as 50 neutrons can be produced by a single high-energy (500 MeV) deuteron. The neutrons could be captured in isotopes such as uranium-238 or thorium-232 to produce new nuclear fuels as discussed in Section 7.3. It has also been suggested that partially burned nuclear fuel can be exposed to neutrons from an accelerator target to bring the fissile isotope content up to operable level.

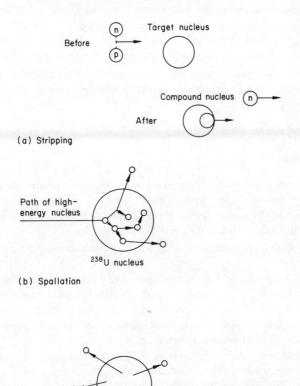

(a) Stripping

(b) Spallation

(c) Evaporation

Fig. 9.9.   Nuclear reactions produced by very high-energy charged particles.

9.7 SUMMARY

Charged particles such as electrons and ions of light elements are brought to high speed and energy by particle accelerators, which employ electric and magnetic fields in various ways. In the high-voltage machines a beam of ions is accelerated directly through a large potential difference, produced by special voltage multiplier circuits or by carrying charge to a positive electrode; in the linear accelerator, ions are given successive accelerations in gaps between tubes lined up in a row; in the cyclotron, the ions are similarly accelerated but move in circular orbits because of the applied magnetic field; in the betatron, a changing magnetic field produces an electric field that accelerates electrons to relativistic speeds. High-energy nuclear physics research is carried out through the use of such accelerators. Through several nuclear processes, high energy charged particles can produce large numbers of neutrons which can create new fissile materials for use as fuels.

9.7   PROBLEMS

9.1. Calculate the potential difference required to accelerate an electron to speed 2×10^5 m/sec.

9.2. What is the proper frequency for a voltage supply to a linear accelerator if the speed of protons in a tube of 0.6 m length is 3×10^6 m/sec?

9.3. Find the time for one revolution of a deuteron in a uniform magnetic field of 1 Wb/m^2.

9.4. Develop a working formula for the final energy of cyclotron ions of mass m, charge q, exit radius R, in a magnetic field B. (Use nonrelativistic energy relations.)

9.5. What magnetic field strength (Wb/m^2) is required to accelerate deuterons in a cyclotron of radius 2.5 m to energy 5 MeV?

9.6. What is the factor by which the mass is increased and what fraction of the speed of light do protons of 200 billion-electron-volts have?

9.7. Calculate the steady deuteron beam current and the electric power required in a 500-GeV accelerator that produces 4 kg per day of plutonium-239. Assume a conservative 25 neutrons per deuteron.

10

Isotope Separators

All of our technology is based on materials in various forms—compounds, alloys, and mixtures. Examples that immediately come to mind are copper for conduction of electricity, steel and concrete for building construction, drugs for medical treatment, and gasoline for propulsion of automobiles. Materials that are mixtures may be prepared by heating and mechanical action; those that are pure elements or compounds may be produced by chemical processing. For several materials used in the nuclear field, however, individual isotopes of elements or a specified combination of isotopes are required. Two important examples are $^{235}_{92}U$ and $^{2}_{1}H$. Since isotopes of a given element have the same atomic number Z, they are essentially identical chemically and thus a physical method that distinguishes between particles on the basis of mass number A is required. In this chapter we shall describe three devices by which uranium isotopes are separated. One is based on differences in ion motion in a magnetic field; the other two on differences in the diffusion of particles through a membrane or against a centrifugal force. Calculations on the amounts of material that must be processed to obtain nuclear fuel will be presented, and estimates of costs given.

10.1 MASS SPECTROGRAPH

We recall from Chapter 9 that a particle of mass m, charge q, and speed v will move in a circular path of radius r if injected perpendicular to a magnetic field of strength B, according to the relation $r = mv/qB$. In the mass spectrograph (Fig. 10.1), ions of the element whose isotopes are to

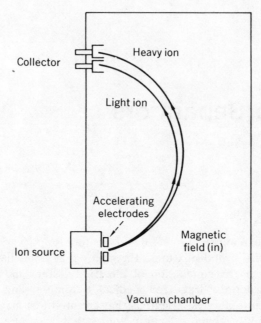

Fig. 10.1. Mass spectrograph.

be separated are produced in an electrical discharge and accelerated through a potential difference V to provide a kinetic energy $\frac{1}{2}mv^2 = qV$. The charges move freely in a chamber maintained at very low gas pressure, guided in semicircular paths by the magnetic field. The heavier ions have a larger radius of motion than the light ions, and the two may be collected separately. It is found (see Problem 10.1) that the distance between the points at which ions are collected is proportional to the difference in the square roots of the masses. The spectrograph can be used to measure masses with some accuracy, or to determine the relative abundance of isotopes in a sample, or to enrich an element in a certain desired isotope.

The electromagnetic process is especially useful for separating light isotopes and those for which small quantities are needed. However, since a large amount of electrical energy is required to provide the magnetic field and the ion acceleration, the cost of large-scale uranium isotope separation by the method is prohibitive, and an alternate process, gaseous diffusion, is the principal one employed to provide reactor fuels.

10.2 GASEOUS DIFFUSION SEPARATOR

The principle of this process can be illustrated by a simple experiment, Fig. 10.2. A container is divided into two parts by a porous membrane and air is introduced on both sides. Recall that air is a mixture of 80% nitrogen, $A = 14$, and 20% oxygen, $A = 16$. If the pressure on one side is raised, the relative proportion of nitrogen on the other side increases. The separation effect can be explained on the basis of particle speeds. The average kinetic energies of the heavy (H) and light (L) molecules in the gas mixture are the same, $E_H = E_L$, but since the masses are different, the typical particle speeds bear a ratio

$$\frac{v_L}{v_H} = \sqrt{\frac{m_H}{m_L}}.$$

Now the number of molecules of a given type that hit the membrane each second is proportional to nv, in analogy to neutron motion discussed in Chapter 5. Those with higher speed thus have a higher probability of passing through the holes in the porous membrane, called the "barrier."

The physical arrangement of one processing unit of a gaseous diffusion plant for the separation of uranium isotopes U-235 and U-238 is shown in

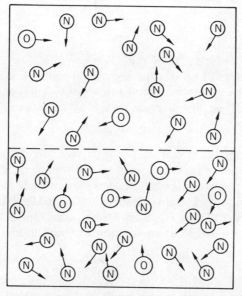

Fig. 10.2. Gaseous diffusion separation of nitrogen and oxygen.

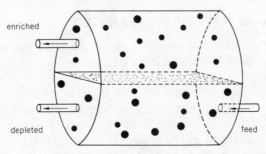

Fig. 10.3.   Gaseous diffusion stage.

Fig. 10.3. A thin nickel alloy serves as the barrier material. In this "stage," gas in the form of the compound uranium hexafluoride (UF_6) is pumped in as feed and removed as two streams. One is enriched and one depleted in the compound $^{235}UF_6$, with corresponding changes in $^{238}UF_6$. Because of the very small mass difference of particles of molecular weight 349 and 352 the amount of separation is small and many stages in series are required in what is called a cascade.

Any isotope separation process causes a change in the relative numbers of molecules of the two species. Let n_H and n_L be the number of molecules in a sample of gas. Their *abundance ratio* is defined as

$$R = \frac{n_L}{n_H}.$$

For example, in ordinary air $R = 80/20 = 4$.

The effectiveness of an isotope separation process is dependent on a quantity called the separation factor r. If we supply gas at one abundance ratio R, the ratio R' on the low-pressure side of the barrier is given by

$$R' = rR.$$

If only a very small amount of gas is allowed to diffuse through the barrier, the separation factor is given by $r = \sqrt{m_H/m_L}$, which for UF_6 is 1.0043. However, for a more practical case, in which half the gas goes through, the separation factor is smaller, 1.0030 (see Problem 10.2). Let us calculate the effect of one stage on natural uranium, 0.711% by weight, corresponding to a U-235 atom fraction of 0.00720, and an abundance ratio of 0.00725. Now

$$R' = rR = (1.0030)(0.00725) = 0.00727.$$

The amount of enrichment is very small. By processing the gas in a series

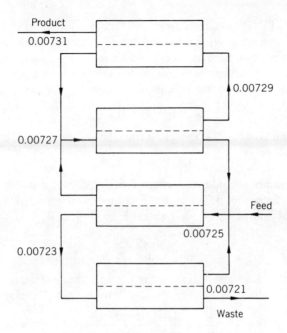

Fig. 10.4.   Gaseous diffusion cascade.

of s stages, each one of which provides a factor r, the abundance ratio is increased by a factor r^s. If R_f and R_p refer to feed and product, respectively, $R_p = r^s R_f$. For $r = 1.0030$ we can easily show that 2375 enriching stages are needed to go from $R_f = 0.00725$ to highly enriched 90% U-235, i.e., $R_p = 0.9/(1 - 0.9) = 9$. Figure 10.4 shows the arrangement of several stages in an elementary cascade, and indicates the value of R at various points. The feed is natural uranium, the product is enriched in U-235, and the waste is depleted in U-235.

10.3 URANIUM ENRICHMENT COSTS

A gaseous diffusion plant is very expensive, of the order of a billion dollars, because of the size and number of components such as separators, pumps, valves, and controls, but the process is basically simple. The plant runs continuously with few operating personnel. The principal operating cost is for the electrical power to provide the pressure differences and perform work on the gas.

The flow of UF_6 and thus uranium through individual stages or the whole plant can be analyzed by the use of material balances. Since the plant operates continuously, one could use atomic mass units but the kilogram per day is most convenient. If the masses of uranium that flow per day through a plant are labeled F, P, and W (for feed, product, and waste), then

$$F = P + W.$$

Letting x stand for the U-235 weight fractions in the flows, the balance for the light isotope is

$$x_f F = x_p P + x_w W.$$

(A similar equation could be written for U-238, but it would contain no additional information.) The two equations can be solved to obtain the ratio of feed and product mass rates. Eliminating W,

$$\frac{F}{P} = \frac{x_p - x_w}{x_f - x_w}.$$

For example, let us find the required feed of natural uranium to obtain 1 kg/day of product containing 3% U-235 by weight, if the waste is 0.2% U-235. Now

$$\frac{F}{P} = \frac{0.03 - 0.002}{0.00711 - 0.002} = 5.5$$

and thus the feed is 5.5 kg/day. We note that W is 4.5 kg/day, which shows that large amounts of depleted uranium "tails" must be stored for each kilogram of U-235 produced. The U-235 content of the tails is too low for use in conventional reactors, but the breeder reactor can convert the U-238 into plutonium, as will be discussed in Chapter 15.

As expected, the higher the enrichment, the greater is the cost of the uranium product. Table 10.1, column 4, shows that the cost if purchased outright ranges from $65/kg to more than $34,000/kg over the range from natural to highly enriched U. Column 2 shows the ratio feed/product, where the figure 5.479 for 3% corresponds to our calculated 5.5. Column 3 is used to calculate the cost of performing enrichment on uranium supplied by a customer. The "separative work" is proportional to the energy required for each kilogram of uranium product handled, and a cost of each separative work unit (SWU) is set by the United States Department of Energy on the basis of current expenses of production.

Suppose that a utility company wants uranium at 3% enrichment for use in its nuclear reactor, and supplies the necessary natural uranium. By use of Table 10.1, the feed required is 5.479 kg for each kilogram of

Table 10.1. Nuclear Fuel and Enrichment Costs

Weight percent U-235	Feed/product	Separative work units (SWU)	Cost of enriched U $/kg†
0.2	——	——	——
0.5	0.587	− 0.173	20.86
0.711	1.000	0	65.00
0.8	1.174	0.104	86.71
1.0	1.566	0.380	139.79
2.0	3.523	2.194	448.40
3.0	5.479	4.306	786.74
5.0	9.393	8.851	1495.65
10.0	19.178	20.863	3332.87
90.0	175.734	227.341	34,156.81

†Assuming $100 per SWU and $65 per kg of natural U ($25/lb U_3O_8).

product. The separative work is 4.306, with cost $430.60 per kilogram. If the utility company were also to return fuel that has been used in a reactor and remains slightly enriched, a credit would be received since the amounts of natural uranium feed and separative work are smaller, as illustrated in Problem 10.6.

10.4 GAS CENTRIFUGE

This device for separating isotopes, also called the ultra-centrifuge because of the very high speeds involved, has been known since the 1940s, but only recently has become popular as a promising alternative to gaseous diffusion. It consists of a cylindrical chamber—the rotor—turning at very high speed in a vacuum (see Fig. 10.5a).

The rotor is driven and supported magnetically. Gas is supplied and centrifugal force tends to compress it in the outer region, but thermal agitation tends to redistribute the gas molecules throughout the whole volume. Light molecules are favored in this effect, and their concentration is higher near the center axis. By various means, a gas flow is established that tends to carry the heavy and light isotopes to opposite ends of the rotor. Depleted and enriched streams of gas are withdrawn, as sketched in Fig. 10.5b. Separation factors of 1.1 or better were obtained with centrifuges about a foot long, rotating at a rate such that the rotor

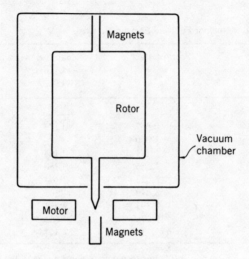

Fig. 10.5a. Gas centrifuge.

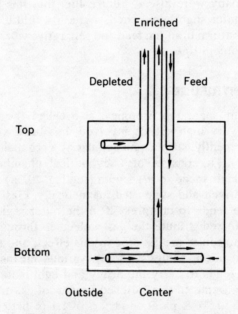

Fig. 10.5b. Gas streams in centrifuge.

surface speed is 350 m/sec. The flow rate per stage of a centrifuge is much lower than that of gaseous diffusion, requiring large numbers of units in parallel. The electrical power consumption for a given capacity is lower, however, by a factor of six to ten. A gaseous diffusion plant must be very large to be efficient, and its billion dollar cost is a very great investment for one company or even a group of companies. The cost of centrifuge plants is much smaller, and they can be added as needed. The only disadvantage of the centrifuge is the lack of experience with the process. Research and development are underway in the United States, Japan, the United Kingdom, West Germany, and the Netherlands. Fig. 10.6 shows a Dutch plant designed to produce 200,000 kg SWU per year.

Fig. 10.6. Centrifuge enrichment demonstration plant at Almelo, the Netherlands. (courtesy of Urenco Ltd. (with thanks to Simon Rippon).)

10.5   LASER ISOTOPE SEPARATION

Research is being done on a separation process based on the internal molecular structure rather than on mass differences. It uses laser light (see Section 2.3) to cause photochemical reactions that are peculiar to the $^{235}UF_6$ molecule, which has vibrational frequencies that are slightly different from those of $^{238}UF_6$. The extreme brightness of the laser beam permits the dissociation of the desired molecule into another material, which is easily separated chemically. A mixture of tails UF_6 from gaseous diffusion and a carrier gas is irradiated first with infrared laser photons, then with ultraviolet laser photons, giving rise to the compound $^{235}UF_5$ as a powder. The system would be designed to convert depleted U into natural U, which would serve as a feed to other enriching devices. The residual material would be essentially pure uranium-238.

10.6   SEPARATION OF DEUTERIUM

The heavy isotope of hydrogen ^2_1H, deuterium, has two principal nuclear applications: (a) as low-absorption moderator for reactors, especially those using natural uranium, and (b) as a reactant in the fusion process. The differences between the chemical properties of light water and heavy water are slight, but sufficient to permit separation of ^1_1H and ^2_1H by several methods. Among these are *electrolysis*, in which the H_2O tends to be more readily dissociated, *fractional distillation*, which takes advantage of the fact that D_2O has a boiling point about 1°C higher than that of H_2O, and *catalytic exchange*, involving the passage of HD gas through H_2O to produce HDO and light hydrogen gas.

10.7   SUMMARY

The separation of isotopes requires a physical process that depends on mass. In the electromagnetic method, as used in a mass spectrograph, ions to be separated travel on circles of different radii. In the gaseous diffusion process, light molecules of a gas diffuse through a membrane more readily than do heavy molecules. The amount of enrichment in gaseous diffusion depends on the square root of the ratio of the masses and is small per stage, requiring a large number of stages. By the use of material balance equations, the amount of feed can be computed, and by the use of tables of work, costs of enriching uranium for reactor fuel can be found. An alternative promising separation device is the gas centrifuge, in which gases diffuse against the centrifugal forces produced by high speeds of rotation. Laser isotope separation involves the selective excitation of $^{235}UF_6$ molecules by lasers to produce chemical reactions. Several methods of separating deuterium from ordinary hydrogen are available.

10.7 PROBLEMS

10.1. (a) Show that the radius of motion of an ion in a mass spectrograph is given by

$$r = \sqrt{\frac{2mV}{qB^2}}.$$

(b) If the masses of heavy (H) and light (L) ions are m_H and m_L, show that their separation at the plane of collection in a mass spectrograph is proportional to $\sqrt{m_H} - \sqrt{m_L}$.

10.2. The ideal separation factor for a gaseous diffusion stage is

$$r = 1 + 0.693(\sqrt{m_H/m_L} - 1).$$

Compute its value for $^{235}UF_6$ and $^{238}UF_6$, noting that $A = 19$ for fluorine.

10.3. (a) Verify that for particles of masses m_H and m_L the number fraction f_L of the light particle is related to the weight fractions w_H and w_L by

$$f_L = \frac{n_L}{n_L + n_H} = \frac{1}{1 + \dfrac{w_H m_L}{w_L m_H}}.$$

(b) Show that the abundance ratio of numbers of particles is either

$$R = \frac{n_L}{n_H} = \frac{f_L}{1 - f_L} \quad \text{or} \quad \frac{w_L/m_L}{w_H/m_H}.$$

(c) Calculate the number fraction and abundance ratio for uranium metal that is 3% U-235 by weight.

10.4. Find the amount of natural uranium feed (0.711% by weight) required to produce 1 kg/day of highly enriched uranium (90% by weight), if the waste concentration is 0.25% by weight. Assume that the uranium is in the form of UF_6.

10.5. How many enriching stages are required to produce uranium that is 3% by weight, using natural UF_6 feed? Let the waste be 0.2%.

10.6. A reactor receives 3% fuel from a gaseous diffusion plant at a rate of 1 kg/day, and returns 1% fuel at 0.98 kg/day to the plant.
(a) Using Table 10.1, show that the fuel returned corresponds to a "credit" of 1.535 kg/day in feed reduction.
(b) Find the value of natural uranium feed to the gaseous diffusion plant.
(c) Find the credit in separative work for the returned uranium.
(d) Find the net separative work and the fuel enrichment cost.

10.7. The number density of molecules as the result of loss through a barrier can be expressed as $n = n_0 \exp(-cvt)$ where c is a constant, v is the particle speed, and n and n_0 are values before and after an elapsed time t. If half the heavy isotope is allowed to pass through, find the abundance ratio $R'/R = r$ in the enriched gas as a function of the ratio of molecular masses. Test the derived formula for the separation of uranium isotopes.

10.8. Depleted uranium (0.2% U-235) is processed by laser separation to yield natural uranium (0.711%). If the feed rate is 1 kg/day and all of the U-235 goes into the product, what amounts of product and waste are produced per day?

11

Radiation Detectors

Measurement of radiation is required in all facets of nuclear energy—in scientific studies, in the operation of reactors for the production of power, and for protection from radiation hazard. Detectors are used to identify products of nuclear reactions, to measure neutron flux, and to determine the amount of radioisotopes in air or water. The kind of detector employed depends on several factors: the particles to be observed—electrons, gamma rays, neutrons, ions, or combinations of them; the energy of the particles; and the environment in which the detector is to be used. The type of measuring device is chosen for the intended purpose and the accuracy desired.

The demands on a detector are related to what it is we wish to know, which may be one or more of the following: (a) that there is a radiation field present, (b) the number of nuclear particles that strike the detector each second (or over some specified period of time), (c) the type of particles present and perhaps the relative number in a mixed radiation field, (d) the energy of the individual particles, (e) the instant a particle arrives at the detector. In this chapter we shall describe the important features of a few popular types of detectors. Most of them are based on the ionization produced by radiation, with the resulting currents passing through an electrical circuit and displayed on a meter or similar indicator. The number of particles of radiation arriving and noted in a given period is obtained by devices called counters.

11.1 GAS COUNTERS

Picture a gas-filled chamber with a central electrode (anode, electrically positive) and a conducting wall (cathode, negative). They are maintained at different potential, as shown in Fig. 11.1. If a charged particle or gamma ray is allowed to enter the chamber, it will produce a certain amount of ionization in the gas. The resultant positive ions and electrons are attracted toward the negative and positive surfaces, respectively. If the voltage across the tube is low, the charges merely migrate through the gas, they are collected, and a current of short duration (a pulse) passes through the resistor and the meter. More generally, amplifying circuits are required. The number of current pulses is a measure of the number of incident particles that enter the detector, which is designated as an *ionization chamber* when operated in this mode.

If the voltage is then increased sufficiently, electrons produced by the incident radiation through ionization are able to gain enough speed to cause further ionization in the gas. Most of this action occurs near the central electrode, where the electric field is highest. The current pulses are much larger than in the ionization chamber because of the amplification effect. The current is proportional to the original number of electrons produced by the incoming radiation, and the detector is now called a *proportional counter*. One may distinguish between the passage of beta particles and alpha particles, which have widely different ability to ionize. The time for collection is very short, of the order of microseconds.

If the voltage on the tube is raised still higher, a particle or ray of any energy will set off a discharge, in which the secondary charges are so great in number that they dominate the process. The discharge stops of its own accord because of the generation near the anode of positive ions,

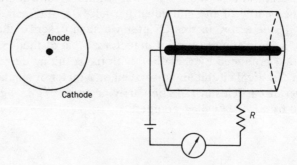

Fig. 11.1. Basic detector.

which reduce the electric field there to such an extent that electrons are not able to cause further ionization. The current pulses are then of the same size, regardless of the event that initiated them. In this mode of operation, the detector is called a *Geiger-Müller (GM) counter*. There is a short period, the "dead time," in which the detector will not count other incoming radiation. If the radiation level is very high, a correction of the observed counts to yield the "true" counts must be made, to account for the dead time. In some gases, such as argon, there is a tendency for the electric discharge to be sustained, and it is necessary to include a small amount of foreign gas or vapor, e.g., alcohol, to "quench" the discharge. The added molecules affect the production of photons and resultant ionization by them.

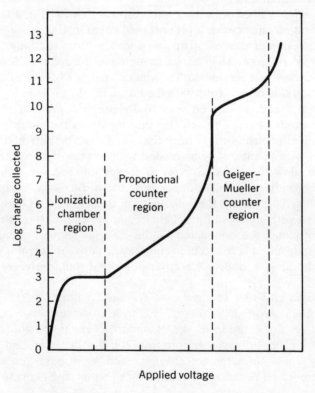

Fig. 11.2. Collection of charge in counters. (From Raymond L. Murray, *Introduction to Nuclear Engineering*, 2nd Ed. © 1961. Reprinted by permission of Prentice-Hall, Inc., Englewood Cliffs, New Jersey.)

A qualitative distinction between the above three types of counters is displayed graphically in Fig. 11.2, which is a semilog plot of the charge collected as a function of voltage. We note that the current varies over several orders of magnitude.

11.2 NEUTRON DETECTORS

In order to detect neutrons, which do not create ionization directly, it is necessary to provide a means for generating the charges that can ionize a gas. Advantage is taken of the nuclear reaction involving neutron absorption in boron

$$_0^1n + {}_5^{10}B \rightarrow {}_2^4He + {}_3^7Li,$$

where the helium and lithium atoms are released as ions. One form of *boron counter* is filled with the gas boron trifluoride (BF_3), and operated as an ionization chamber or a proportional counter. It is especially useful for the detection of thermal neutrons since the cross section of boron-10 at 0.0253 eV is large, 3838 barns, as noted in Chapter 5. Most of the 2.8 MeV energy release goes to the kinetic energy of the product nuclei. The reaction rate of neutrons with the boron in BF_3 gas is independent of the neutron speed, as can be seen by forming the product $R = nvN\sigma_a$, where σ_a varies as $1/v$. The detector thus measures the number density n of an incident neutron beam rather than the flux. Alternatively, the metal electrodes of a counter may be coated with a layer of boron that is thin enough to allow the alpha particles to escape into the gas. The counting rate in a boron-lined chamber depends on the surface area exposed to the neutron flux. To enhance the sensitivity of detection, the counter can be constructed with a series of parallel circular plates (PCP) on a rod, while alternating plates are connected to the case, as shown in Fig. 11.3. A similar arrangement is found in gamma ray sensitive detectors, where the gammas produce secondary electrons in a metal wall, preferably of large Z value.

The *fission chamber* is often used for slow neutron detection. A thin layer of U-235, with high thermal neutron cross section, 678 barns, is deposited on the cathode of the chamber. Energetic fission fragments produced by neutron absorption traverse the detector and give the necessary ionization. Uranium-238 is avoided because it is not fissile with slow neutrons and because of its stopping effect on fragments from U-235 fission.

Neutrons in the thermal range can be detected by the radioactivity induced in a substance in the form of small foil or thin wire. Examples are

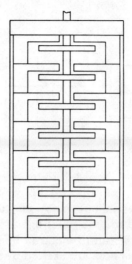

Fig. 11.3.   Parallel circular plate ionization chamber.

manganese $^{55}_{25}$Mn, with a 13.3 barn cross section at 2200 m/sec, which becomes $^{56}_{25}$Mn with half-life 2.58 h; and dysprosium $^{164}_{66}$Dy, 900 barns, becoming $^{165}_{66}$Dy, half-life 140 min. For detection of neutrons slightly above thermal energy, materials with a high resonance cross section are used, e.g., indium, with a peak at 1.45 eV. To separate the effects of thermal neutron capture and resonance capture, comparisons are made between measurements made with thin foils of indium and those of indium covered with cadmium. The latter screens out low-energy neutrons (below 0.5 eV) and passes those of higher energy.

For the detection of fast neutrons, up in the MeV range, the *proton recoil* method is used. We recall from Chapter 5 that the scattering of a neutron by hydrogen results in an energy loss, which is an energy gain for the proton. Thus a hydrogenous material such as methane (CH_4) or H_2 itself may serve as the counter gas. The energetic protons play the same role as did alpha particles and fission fragments in the counters discussed previously. Nuclear reactions such as $^3_2$He (n, p) $^3_1$H can also be employed to obtain detectable charged particles.

11.3 SCINTILLATION COUNTERS

The name of this detector comes from the fact that the interaction of a particle with some materials gives rise to a scintillation or flash of light.

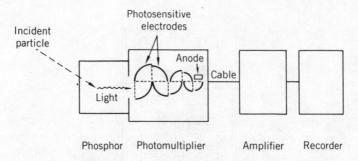

Fig. 11.4. Scintillation detection system.

The basic phenomenon is familiar—many substances can be stimulated to glow visibly on exposure to ultraviolet light, and the images on a color television screen are the result of electron bombardment. Molecules of materials classed as phosphors are excited by radiation such as charged particles and subsequently emit pulses of light. The substances used in the scintillation detector are inorganic, e.g., sodium iodide or lithium iodide, or organic, in one of various forms—crystalline, plastic, liquid, or gas.

The amount of light released when a particle strikes a phosphor is proportional to the energy deposited, and thus makes the detector especially useful for the determination of particle energies. Since charged particles have a short range, most of their energy appears in the substance. Gamma rays also give rise to an energy deposition through electron recoil in both the photoelectric effect and Compton scattering, and through the pair production–annihilation process. A schematic diagram of a detector system is shown in Fig. 11.4. Some of the light released in the phosphor is collected in the photomultiplier tube, which consists of a set of electrodes with photosensitive surfaces. When a photon strikes the surface, an electron is emitted by the photoelectric effect, it is accelerated to the next surface where it dislodges more electrons, and so on, and a multiplication of current is achieved. An amplifier then increases the electrical signal to a level convenient for counting or recording.

Radiation workers are required to wear personal detectors called dosimeters in order to determine the amount of exposure to X- or gamma rays or neutrons. Among the most reliable and accurate types is the thermoluminescent dosimeter, which measures the energy of radiation absorbed. It contains crystalline materials such as CaF_2 or LiF which store energy in excited states of the lattice called traps. When the substance is heated, it releases light in a typical "glow curve" as shown in Fig. 11.5. The dosimeter consists of a small vacuum tube with a coated cylinder that can be heated by a built-in filament when the tube is plugged into a voltage supply. A photomultiplier reads the peak of the glow curve and gives values of the accumulated energy absorbed, i.e., the dose. The device is linear in its response over a very wide range of exposures; it can be used over and over with little change in behavior.

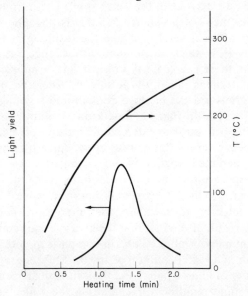

Fig. 11.5.   Glow curve of the phosphor CaF_2.

11.4 SOLID-STATE DETECTORS

The use of a solid rather than a gas in a detector has the advantage of compactness, since the range of nuclear particles is very short in solids. In addition, when the detector is composed of semiconductor materials, great accuracy in the measurement of energy and time of arrival is achieved. There are many similarities between gas and solid detectors, but the mechanism of ion motion is quite different. We can visualize a crystal semiconductor, such as silicon or germanium, as a regular array of fixed atoms, with some freedom of electron motion in the lattice. An incident charged particle can detach an electron from an atom and cause it to leave the vicinity. The removal of the electron causes a vacancy or "hole," which acts effectively as a positive charge. The electron-hole *pair* produced is analogous to negative and positive ions in a gas. Electrons can migrate through the material or be carried along by an electric field, while the holes "move" as electrons are successively exchanged with neighboring atoms.

The electrical conductivity of a semiconductor is very sensitive to the presence of certain impurities. Consider silicon, which has a chemical valence of 4, i.e., 4 electrons in the outer shell. The introduction of small amounts of an element with valence 5, such as phosphorous or arsenic, provides additional negative charge, and the resulting material is classed as *n*-type silicon. If, instead, an element with valence 3 is added, such as boron or gallium, there is a scarcity of electrons or an excess of positive holes, and the material is called *p*-type silicon. When two layers of *n*-type and *p*-type materials are put in close contact and a voltage is applied to the outside surfaces as in Fig. 11.6, electrons are drawn one way, holes the other, leaving a neutral or "depleted" region between. Most of the voltage drop occurs across the neutral zone, since it is nearly a perfect insulator. The depleted region is very sensitive to radiation. The electron-hole pairs resulting from an incident particle are swept out by the high internal electric field and register as a current pulse. The ability of an *n*–*p* junction detector to measure the energy of nuclear particles accurately is the result of the fact that the energy required to create a pair in silicon or germanium is about 3 eV (in comparison with about 32 eV to create an ion pair in a gas). A photon of say 0.1 MeV energy gives rise to a very large number of charge pairs, and thus statistical accuracy is higher. In addition, the time required for charges to be collected is extremely short, about one billionth of a second, permitting precise measurement of the time of the counting event.

The introduction of the element lithium to a crystal of silicon or germanium dramatically increases the depth of the depletion region by compensating for the residual fixed charges. Improved resolution of the gamma rays from nuclei is thus achieved. Such detectors, designated as Ge(Li), i.e., lithium-drifted germanium, have depletion depths of up to 3 cm and useful volumes of 100 cm^3. They must be maintained at all times at liquid nitrogen temperature, $-195.8°C$. The same benefit has been achieved recently by the advent of germanium of exceedingly high purity (one impurity atom per 10^{13} atoms).

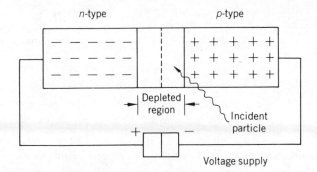

Fig. 11.6. Solid-state *n-p* junction detector.

11.5 SUMMARY

The detection of radiation and the measurement of its properties is required in all aspects of the nuclear field. In gas counters, the ionization produced by incoming radiation is collected. Dependent on the voltage applied between electrodes, counters detect all particles or distinguish between types. Neutrons are detected indirectly by the products of nuclear reactions—for slow neutrons by absorption in boron or uranium, for fast neutrons by scattering in hydrogen. Scintillation counters release measurable light on bombardment by charged particles or gamma rays, while solid-state detectors take advantage of the sensitivity of semiconductors to a disturbance of the charge balance.

11.6 PROBLEMS

11.1. (a) Find the number density of molecules of BF_3 in a detector of 2.5 cm diameter to be sure that 90% of the thermal neutrons incident along a diameter are caught (σ_a for natural boron is 760 barns).
(b) How does this compare with the number density for the gas at atmospheric pressure, with density 3.0×10^{-3} g/cm³?
(c) Suggest ways to achieve the high efficiency desired.

11.2. An incident particle ionizes helium to produce two electrons and an alpha particle halfway between two parallel plates with potential difference between them. If the gas pressure is very low, estimate the ratio of the times elapsed until the charges are collected, t_e/t_α. Discuss the effect of collisions on the collection time.

11.3. We collect a sample of gas suspected of containing a small amount of radioiodine, half-life 8 days. If we observe in a period of 1 day a total count of 50,000 in a counter that detects all radiation emitted, how many atoms were initially present?

11.4. In a gas counter, the potential difference at any point r between a central wire of radius r_1 and a concentric wall of radius r_2 is given by

$$V = V_0 \frac{\ln (r/r_1)}{\ln (r_2/r_1)},$$

where V_0 is the voltage across the tube. If $r_1 = 1$ mm and $r_2 = 1$ cm, what fraction of the potential difference exists within a millimeter of the wire?

11.5. How many electrodes would be required in a photomultiplier tube to achieve a multiplication of one million if one electron releases four electrons?

12

Neutron Chain Reactions

The possibility of a chain reaction involving neutrons and a nuclear fuel such as uranium is dependent on the number of neutrons produced per absorption, η, as discussed in Chapter 7. Its value must be more than one because of inevitable losses of neutrons. However, to achieve a self-sustaining chain reaction, one in which no neutrons need to be supplied, a certain amount of uranium must be brought together, the "critical" mass. In order to appreciate this requirement, we can visualize an experiment in which we can assemble various amounts of U-235. In effect, we will be building a nuclear reactor but will not consider all of the materials and components. We ignore parts that are present for structural support, to provide control, to permit the extraction of energy, or to give protection from radiation. The minimum ingredients are a nuclear fuel and at least one neutron to start the process.

12.1 THE SELF-SUSTAINING CHAIN REACTION

If we have only one atom of U-235, and it is bombarded by a neutron to induce fission, the resultant neutrons are able to do nothing further, there being no more fuel. Instead, we form a small sphere of uranium of volume, say, $1 \, cm^3$ containing about 19 g. The number of nuclei is adequate for a very long sequence of fission events, but on introducing a neutron, the series of reactions soon ends because of loss from the surface of the sphere. Only if we were to supply neutrons continually to make up for this "leakage" could we keep the reaction going. Such an assembly is called "subcritical." However, if we bring together about 50 kg of U-235 metal in spherical form, the rate of neutron production is

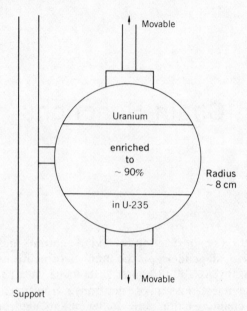

Fig. 12.1. Fast metal assembly "Godiva."

sufficient to balance leakage losses, and an outside supply of neutrons is not needed. In terms of neutrons, the assembly is self-sustaining or "critical." The amount of nuclear fuel is called the critical mass, its volume is the critical size. Figure 12.1 shows the uranium metal assembly Godiva, so named because it was "bare," i.e., had no surrounding materials. This reactor, composed of highly enriched uranium, was used for many years for test purposes at Los Alamos, New Mexico.

If we add still more uranium beyond the 50 kg required for criticality, more neutrons are produced than are used, the neutron population grows, and the reactor is "supercritical." Early nuclear weapons consisted of two hemispheres of uranium, each subcritical and unable separately to sustain a chain reaction. When suddenly brought together they formed a highly supercritical mass, in which the rapid growth in the numbers of neutrons, fission events, and energy developed produced a violent explosion.

12.2 MULTIPLICATION FACTORS

We may describe any arrangement of fuel material by a single number, the effective multiplication factor k (or k_{eff}), as being the *net* number of neutrons produced (accounting for all possible losses) per initial neutron.

If k is less than 1, the system is subcritical; if k is equal to 1, it is critical; and if k is greater than 1, supercritical. The design and operation of all reactors is focused on k or on related quantities.† The choice of materials and size is made to assure a balance between neutron production by fission and losses by capture in other elements or leakage from the boundaries of the assembly. In the process of bringing parts together in what is called a critical experiment, observations are made that give estimates of k. During operation, variations in k are made as needed by adjustments of neutron-absorbing rods or dispersed chemicals that cause increases or decreases in the neutron population. Eventually, in the operation of a reactor for a long time, enough fuel is consumed that k goes below 1 regardless of adjustments of control materials, and the reactor must be shut down for refueling.

Two views of the growth of human population are analogous to neutron multiplication. A person born today in the United States has a life expectancy of about 70 yr, which is a statistical result of past data on individual longevity. Alternatively, we may say that the birthrate exceeds the death rate such that there is a population growth of 2% per year. The first view involves the probability of survival or death of the individual; the second view compares rates of competing processes that affect the total population.

Similarly, we can focus attention on a typical neutron that starts its life in fission, and has various chances of dropping out of the cycle because of leakage and absorption in other materials besides fuel. Or, at a given time we can form the sum of the reaction rates for the processes of neutron absorption, fission with neutron yield, and leakage in order to find out if the neutron population is increasing, is steady, or is decreasing. Each method has its merits for purposes of discussion or analysis.

The statistical approach involves the observation of many histories and deducing averages. Let us look at the possible behaviors of several fission neutrons, using the uranium metal reactor for reference. As in Fig. 12.2(a), a neutron may escape on first flight from the sphere, since mean free paths are rather long. Another (b) may make a scattering collision and then escape. Others may collide and be absorbed either (c) to form U-236 or (d) to give rise to fission, the latter case yielding three neutrons in the case shown. Several collisions may occur before leakage or absorption takes place. A "flow diagram" as in Fig. 12.3 is useful to describe the fates. The boxes represent processes, the circles the numbers of neutrons.

†Such as $\delta k = k - 1$ or $\delta k/k$ called reactivity, symbolized by ρ.

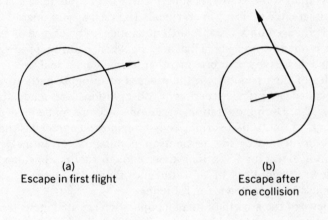

<table>
<tr><td>(a)
Escape in first flight</td><td>(b)
Escape after
one collision</td></tr>
</table>

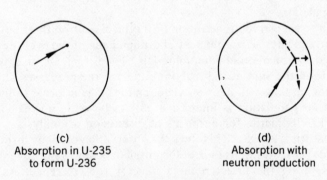

(c)
Absorption in U-235
to form U-236

(d)
Absorption with
neutron production

Fig. 12.2. Neutron histories.

The fractions of absorbed neutrons that form U-236 and that cause fission, respectively, are the ratios of the cross section for capture σ_c and fission σ_f to that for absorption σ_a. The average number of neutrons produced by fission is ν, where for simplicity we omit the bar signifying average. Now let η be the combination $\nu\sigma_f/\sigma_a$, and note that it is the number of neutrons per absorption in uranium. Thus letting $\mathscr{L}$ be the

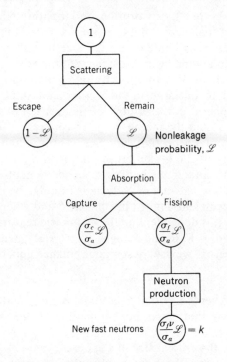

Fig. 12.3.   Neutron cycle in metal assembly.

fraction *not* escaping by leakage,

$$k = \eta \mathscr{L}.$$

The system is critical if $k = 1$, or $\eta \mathscr{L} = 1$. Measurements show that η is around 2.2 for fast neutrons, thus $\mathscr{L}$ must be $1/2.2 = 0.45$, which says that as many as 45% of the neutrons must remain in the sphere, while no more than 55% escape through its boundary.

12.3 NEUTRON BALANCES

Let us now look at the same uranium metal system from the other viewpoint, that of reaction rates. A critical reactor is one in which there is a balance of processes:

$$\text{production} = \text{absorption} + \text{leakage}.$$

This word equation relates the rates of neutron generation by fission, of removal from the system by absorption, and loss through the boundaries. The statement of balance can be put in a form that expresses the relationship of materials content and the size and shape of the reactor, according to the following logic. Suppose that the reactor is pictured as a single region of volume V with no surrounding material. The neutron flux may be expected to be large in the center and small near the edges because of the general outward flow of neutrons. Let the average flux over the core be $\bar{\phi}$ and the macroscopic cross section for absorption be Σ_a. Then the rate of *absorption* is $\bar{\phi}\Sigma_a V$. The production term can be formed by analogy. If the fission cross section is Σ_f, the fission rate over the core is $\bar{\phi}\Sigma_f V$, and if each fission yields ν neutrons, the rate of *production* is $\bar{\phi}\Sigma_f V\nu$. The rate of *leakage* is found to be $\bar{\phi}DB^2V$, where D is the diffusion coefficient defined in Chapter 5 and where a new quantity B^2, called *buckling*, is dependent on the geometric features of the reactor. For a sphere of radius R, the buckling is $B^2 = (\pi/R)^2$, with other formulas for various shapes.† If we use these in our balance equation and simplify, we find

$$\Sigma_f\nu = \Sigma_a + DB^2.$$

This formula represents a *critical condition*, relating material content and geometric features that must be compatible if the system is to be self-sustaining.

Let us calculate the critical size of a sphere of pure U-235. For metal of density $19\,\text{g/cm}^3$, $N_u = 0.048\,\text{cm}^{-3}$ (in units of 10^{24}), and with cross sections $\sigma_f = 1.40$ barns, $\sigma_a = 1.65$ barns, $\sigma_t = 6.8$ barns, and $\nu = 2.6$, we find $\Sigma_f = 0.0672\,\text{cm}^{-1}$, $\Sigma_a = 0.0792\,\text{cm}^{-1}$, $\Sigma_t = 0.326\,\text{cm}^{-1}$, $\lambda_t = 1/\Sigma_t = 3.07\,\text{cm}$, $D = \lambda_t/3 = 1.02\,\text{cm}$, $B^2 = (\Sigma_f\nu - \Sigma_a)/D = 0.0936\,\text{cm}^{-2}$, and $R = \pi/B = 10.3\,\text{cm}$. A slight correction is required to take proper account of the shape of the flux near the boundary. The calculated radius is smaller by an amount $0.71\,\lambda_t$, or 2.2 cm for this case. We thus predict the sphere radius to be 8.1 cm, which is in reasonable agreement with the actual size of the Godiva reactor. The core volume is $V = \frac{4}{3}\pi R^3 = 2.2 \times 10^3\,\text{cm}^3$, and using the given density, the calculated critical mass is 42 kg.

The critical condition $\Sigma_f\nu = \Sigma_a + DB^2$ enables us to make a variety of reactor calculations. If the size were known, the necessary properties of fuel can be found, the reverse of the case just examined. As another example, suppose we substitute a fuel that releases more neutrons in fission. To keep the reactor critical, the value of $B^2 = (\pi/R)^2$ must be

†For a rectangular parallelepiped of sides L, W, and H, it is $B^2 = (\pi/L)^2 + (\pi/W)^2 + (\pi/H)^2$; thus for a cube of side s, it is $B^2 = 3(\pi/s)^2$; for a circular cylinder of height H and radius R, $B^2 = (2.405/R)^2 + (\pi/H)^2$.

raised, by decreasing the radius R. If we add a strong absorber such as boron, Σ_a will increase, B^2 must be smaller, and thus R must increase to compensate. The reader can investigate the effect of uniform expansion which reduces the number density of all materials and also causes the radius to change (see Problem 12.3). It is easy to show that our critical condition is identical to $1 = \eta\mathscr{L}$ if $\eta = (\Sigma_f \nu / \Sigma_a)_u$ and $\mathscr{L} = 1/(1 + DB^2/\Sigma_a)$.

12.4 REACTOR POWER

The power developed by a reactor is a quantity of great interest for practical reasons. Power is related to the neutron population, and also to the mass of fissile material present. First, let us look at a typical cubic centimeter of the reactor, containing N fuel nuclei, each with cross section for fission σ_f at the typical neutron energy of the reactor, corresponding to neutron speed v. Suppose that there are n neutrons in the volume. The rate at which the fission reaction occurs is thus $R_f = nvN\sigma_f$ fissions per second. If each fission produces an energy w, then the power per unit volume is $p = wR_f$. For the whole reactor, of volume V, the rate of production of thermal energy is $P = pV$. If we have an average flux $\bar{\phi} = nv$ and a total number of fuel atoms $N_T = NV$, the total reactor power is seen to be

$$P = \bar{\phi}N_T\sigma_f w.$$

Thus we see that the power is dependent on the product of the number of neutrons and the number of fuel atoms. A high flux is required if the reactor contains a small amount of fuel, and conversely. All other things equal, a reactor with a high fission cross section can produce a required power with less fuel than one with small σ_f. We recall that σ_f decreases with increasing neutron energy. Thus for given power P, a "fast" reactor, one operating with neutron energies principally in the vicinity of 1 MeV, requires either a much larger flux or a larger fissile fuel mass than does the "thermal" reactor, with neutrons of energy around 0.1 eV.

The power developed by most familiar devices is closely related to fuel consumption. For example, a large car generally has a higher gasoline consumption rate than a small car, and more gasoline is used in operation at high speed than at low speed. In a reactor, it is necessary to add fuel very infrequently because of the very large energy yield per pound, and the fuel content remains essentially constant. From the formula relating power, flux, and fuel, we see that the power can be readily raised or lowered by changing the flux. By manipulation of control rods, the neutron population is allowed to increase or decrease to the proper level.

Power reactors used to generate electricity produce about 3000 megawatts of thermal power (MWt), and with an efficiency of conversion of around $\frac{1}{3}$, give 1000 MW of electrical power (1000 MWe).

12.5   MULTIPLICATION IN A THERMAL REACTOR

The presence of large amounts of neutron-moderating material such as water in a reactor greatly changes the neutron distribution in energy. Fast neutrons slow down by means of collisions with light nuclei, with the result that most of the fissions are produced by low-energy (thermal) neutrons. Such a system is called a "thermal" reactor in contrast with a system without moderator, a "fast" reactor, operating principally with fast neutrons. The cross sections for the two energy ranges are widely different, as noted in Problem 12.6. Also, the neutrons are subject to being removed from the multiplication cycle during the slowing process by strong resonance absorption in elements such as U-238. Finally, there is competition for the neutrons between fuel, coolant, structural materials, fission products, and control absorbers.

The description of the multiplication cycle is somewhat more complicated than that for a fast metal assembly, as seen in Fig. 12.4. The set of reactor parameters are (a) the fast fission factor ϵ, representing the immediate multiplication because of fission at high neutron energy, mainly in U-238; (b) the fast nonleakage probability $\mathscr{L}_f$, being the fraction remaining in the core during neutron slowing; (c) the resonance escape probability p, the fraction of neutrons *not* captured during slowing; (d) the thermal nonleakage probability $\mathscr{L}_t$, the fraction of neutrons remaining in the core during diffusion at thermal energy, (e) the thermal utilization f, the fraction of thermal neutrons absorbed in fuel; and (f) the reproduction factor η, as the number of new fission neutrons per absorption in fuel. At the end of the cycle starting with one fission neutron, the number of fast neutrons produced is seen to be $\epsilon p f \eta \mathscr{L}_f \mathscr{L}_t$, which may be also labeled k, the effective multiplication factor. It is convenient to group four of the factors to form $k_\infty = \epsilon p f \eta$, the so-called "infinite multiplication factor" which would be identical to k if the medium were infinite in extent, without leakage. If we form a composite nonleakage probability $\mathscr{L} = \mathscr{L}_f \mathscr{L}_t$, then we may write

$$k = k_\infty \mathscr{L}.$$

For a reactor to be critical, k must equal 1, as before.

To provide some appreciation of the sizes of various factors, let us calculate the values of the composite quantities for a thermal

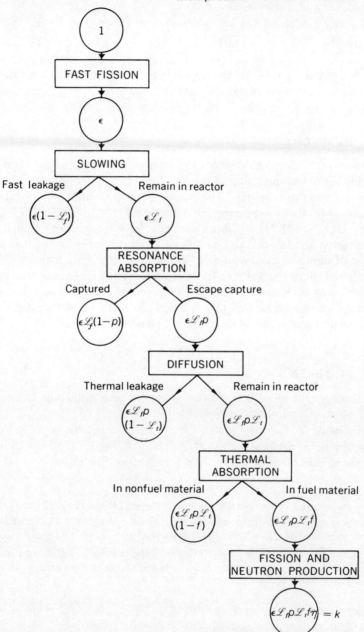

Fig. 12.4.   Neutron cycle in thermal reactor.

reactor, for which $\epsilon = 1.03$, $p = 0.71$, $\mathscr{L}_f = 0.97$, $\mathscr{L}_t = 0.99$, $f = 0.79$, and $\eta = 1.8$. Now $k_\infty = (1.03)(0.71)(1.8)(0.79) = 1.04$, $\mathscr{L} = (0.97)(0.99) = 0.96$, and $k = (1.04)(0.96) = 1.00$. For this example, the various parameters yield a critical system. In the following chapter we shall describe the physical construction of typical thermal reactors.

12.6 SUMMARY

A self-sustaining chain reaction involving neutrons and fission is possible if the accumulation of nuclear fuel is large enough, i.e., a critical mass is brought together. The value of the multiplication factor k indicates whether a reactor is subcritical (< 1), critical ($= 1$) or supercritical (> 1). There is an analogy between neutron and human populations; and in both, the history of a typical individual or the rate of change of numbers can be studied. The critical condition formula permits the calculation of reactor size for given materials content (or vice versa), and the study of effects of changes. The reactor power, dependent on the product of flux and fuel atoms, is readily adjustable. A thermal reactor contains moderator and operates on slowed neutrons.

12.7 PROBLEMS

12.1. Calculate the reproduction factor η for fast neutrons in pure U-235, using values of σ_f, σ_a, and ν cited in the text.

12.2. Calculate $\mathscr{L}$ and show that $\eta \mathscr{L} = 1$ is met numerically by the calculated size of the U-235 metal assembly.

12.3. Using the critical condition for a bare reactor, written as

$$\Sigma_f \nu - \Sigma_a = DB^2,$$

where for a cube $B^2 = 3(\pi/s)^2$, examine the effect on criticality of uniform thermal expansion. The cross sections and diffusion coefficient are dependent on fuel number density N_u, which varies inversely with volume $V = s^3$.

12.4. If the power developed by the Godiva-type reactor is 100 W, and the total mass of fuel is 50 kg, what average flux is implied? Note that the energy per fission is $w \simeq 3.04 \times 10^{-11}$ W-sec.

12.5. Find the multiplication factors k_∞ and k for a thermal reactor for which $\epsilon = 1.05$, $p = 0.75$, $\mathscr{L}_f = 0.90$, $\mathscr{L}_t = 0.98$, $f = 0.85$, and $\eta = 1.75$. Evaluate the reactivity ρ.

12.6. The number of neutrons per absorption in partly enriched uranium depends on the properties of the two isotopes (1) U-235 and (2) U-238 as follows:

$$\eta = \frac{N_1\sigma_{f1}\nu_1 + N_2\sigma_{f2}\nu_2}{N_1\sigma_{a1} + N_2\sigma_{a2}}.$$

Compute the value of η for uranium of 3% U-235 by weight, $N_1/N_2 = 0.0315$, (Problem 10.3) for two cases: (a) a fast reactor, and (b) a thermal reactor, using the table of data below. Comment on the results.

	Thermal	Fast
σ_{f1}	580	1.40
σ_{a1}	678	1.65
σ_{f2}	0	0.095
σ_{a2}	2.70	0.255
ν_1	2.42	2.6
ν_2	0	2.6

12.7. Find the weight percent of U-235 in uranium that will yield $\eta = 1.75$ for thermal neutrons. Suggestion: use formulas from Problems 12.6 and 10.3.

12.8. Calculate η, the number of neutrons per absorption in fuel, for uranium oxide in which the U-235 atom fraction is 0.2, regarded as a practical lower limit for nuclear weapons material. Use thermal neutron data from Problem 12.6.

13

Nuclear Reactor Concepts

Although the only requirement for a neutron chain reaction is a sufficient amount of a fissionable element, many combinations of materials and arrangements can be used to construct an operable nuclear reactor. Several different types or concepts have been devised and tested over the period since 1942, when the first reactor started operation, just as various kinds of engines have been used—steam, internal combustion, reciprocating, rotary, jet, etc. Experience with the individual reactor concepts has led to the selection of a few that are most suitable, using criteria such as economy, reliability, and ability to meet performance demands.

In this chapter we shall identify these important reactor features, compare several concepts, and then focus attention on the components of one specific power reactor type. We shall then examine the processes of fuel consumption and control in a power reactor.

13.1 REACTOR CLASSIFICATION

A general classification scheme for reactors has evolved that is related to the distinguishing features of the reactor types. These features are listed below.

(a) Purpose

The majority of reactors in operation or under construction have as purpose the generation of large blocks of commercial electric power. Others serve training or radiation research needs, and many provide propulsion power for submarines. Available also are tested reactors for commercial surface ships and for spacecraft. At various stages of development of a new concept, such as the breeder reactor, there will be constructed a prototype reactor, one which tests feasibility, and a demonstration reactor, one that evaluates commercial possibilities.

(b) Neutron Energy

A fast reactor is one in which most of the neutrons are in the energy range 0.1–1 MeV, below but near the energy of neutrons released in fission. The neutrons remain at high energy because there is relatively little material present to cause them to slow down. In contrast, the thermal reactor contains a good neutron moderating material, and the bulk of the neutrons have energy in the vicinity of 0.1 eV.

(c) Moderator and Coolant

In some reactors, one substance serves two functions—to assist in neutron slowing and to remove the fission heat. Others involve one material for moderator and another for coolant. The most frequently used materials are listed below:

Moderators	Coolants
light water	light water
heavy water	carbon dioxide
graphite	helium
beryllium	liquid sodium

The condition of the coolant serves as a further identification. The *pressurized water reactor* provides high-temperature water to a heat exchanger that generates steam, while the *boiling water reactor* supplies steam directly.

(d) Fuel

Uranium with U-235 content varying from natural uranium ($\simeq 0.7\%$) to slightly enriched ($\simeq 3\%$) to highly enriched ($\simeq 90\%$) is employed in various reactors, with the enrichment depending upon what other absorbing materials are present. The fissile isotopes $^{239}_{94}\text{Pu}$ and $^{233}_{92}\text{U}$ are produced and consumed in reactors containing significant amounts of U-238 or Th-232. Plutonium serves as fuel for fast breeder reactors and can be recycled as fuel for thermal reactors. The fuel may have various physical forms—a metal, or an alloy with a metal such as aluminum, or a compound such as the oxide UO_2 or carbide UC.

(e) Arrangement

In most modern reactors, the fuel is isolated from the coolant in what is called a *heterogeneous* arrangement. The alternative is a homogeneous mixture of fuel and moderator or fuel and moderator-coolant.

(f) Structural Materials

The functions of support, retention of fission products, and heat conduction are provided by various metals. The main examples are aluminum, stainless steel, and zircaloy, an alloy of zirconium and zinc.

By placing emphasis on one or more of the above features of reactors, reactor concepts are identified. Some of the more widely used or promising power reactor types are the following:

PWR (pressurized water reactor), a thermal reactor with light water at high pressure (2200 psi) and temperature (600°F) serving as moderator-coolant, and a heterogeneous arrangement of slightly enriched uranium fuel.

BWR (boiling water reactor), similar to the PWR except that the pressure and temperature are lower (1000 psi and 550°F).

HTGR (high temperature gas-cooled reactor), using graphite moderator, highly enriched uranium with thorium, and helium coolant (1430°F and 600 psi).

CANDU (Canadian deuterium uranium) using heavy water moderator

Table 13.1. Power Reactor Materials.

	Pressurized water (PWR)	Boiling water (BWR)	Natural uranium heavy water (CANDU)	High temp. gas-cooled (HTGR)	Liquid metal fast breeder (LMFBR)
Fuel form	UO_2	UO_2	UO_2	UC_2, ThC_2	PuO_2, UO_2
Enrichment	3% U-235	2.5% U-235	0.7% U-235	93% U-235	15 wt.% Pu-239
Moderator	water	water	heavy water	graphite	none
Coolant	water	water	heavy water	helium gas	liquid sodium
Cladding	zircaloy	zircaloy	zircaloy	graphite	stainless steel
Control	B_4C or Ag–In–Cd rods	B_4C crosses	moderator level	B_4C rods	tantalum or B_4C rods
Vessel	steel	steel	steel	prestressed concrete	steel

and natural uranium fuel that can be loaded and discharged during operation.

LMFBR (liquid metal fast breeder reactor), with no moderator, liquid sodium coolant, and plutonium fuel, surrounded by natural or depleted uranium.

Table 13.1 amplifies on the principal features of the five main power reactor concepts.

13.2 POWER REACTORS

The large-scale reactors used for the production of thermal energy that is converted to electrical energy are much more complex than the fast assembly described in Chapter 12. To illustrate, we can identify the components and their functions in a modern pressurized water reactor. Figure 13.1 gives some indication of the sizes of the various parts.

The fresh fuel installed in a typical PWR consists of cylindrical pellets of slightly enriched (3% U-235) uranium oxide (UO_2) of diameter about $\frac{3}{8}$in. ($\sim$ 1 cm) and length about 0.6 in. ($\sim$ 1.5 cm). A zircaloy tube of wall thickness 0.025 in. ($\sim$ 0.6 mm) is filled with the pellets to an "active length" of 12 ft (365 cm) and sealed to form a fuel rod (or pin). The metal container serves to provide support for the column of pellets, to provide cladding that retains radioactive fission products, and to protect the fuel from interaction with the coolant. About 200 of the fuel pins are grouped in a bundle called a fuel element of about 8 in. ($\sim$ 20 cm) on a side, and about 180 elements are assembled in an approximately cylindrical array to form the reactor *core*. This structure is mounted on supports in a *steel pressure vessel* of outside diameter about 16 ft ($\sim$5 m), height 40 ft ($\sim$12 m) and walls up to 12 in. ($\sim$30 cm) thick. *Control rods*, consisting of an alloy of cadmium, silver, and indium, provide the ability to change the amount of neutron absorption. The rods are inserted in some vacant fuel pin spaces and magnetically connected to drive mechanisms. On interruption of magnet current, the rods enter the core through the force of gravity. The pressure vessel is filled with light water, which serves as neutron moderator, as coolant to remove fission heat, and as *reflector*, the layer of material surrounding the core that helps prevent neutron escape. The water also contains in solution the compound boric acid, H_3BO_3, which strongly absorbs neutrons in proportion to the number of boron atoms and thus inhibits neutron multiplication, i.e., "poisons" the reactor. The term *soluble poison* is often used to identify this material, the concentration of which can be adjusted during reactor operation. To keep the reactor critical as fuel is consumed, the boron content is gradually reduced. A *shield* of concrete surrounds the pressure vessel and other equipment to provide protection against neutrons and gamma rays from the nuclear reactions. The shield also serves as an additional barrier to the release of radioactive materials.

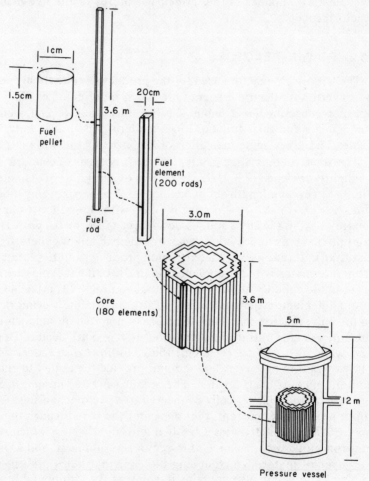

Fig. 13.1.   Reactor construction.

We have mentioned only the main components, which distinguish a nuclear reactor from other heat sources such as one burning coal. An actual system is much more complex than described above. It contains equipment such as spacers to hold the many fuel rods apart; core support structures; baffles to direct coolant flow effectively; guides, seals, and motors for the control rods; guide tubes and electrical leads for neutron-detecting instruments, brought through the bottom of the pressure vessel and up into certain fuel assemblies; and bolts to hold down the vessel head and maintain the high operating pressure.

The power reactor is designed to withstand the effects of high temperature, erosion by moving coolant, and nuclear radiation. The materials of construction are chosen for their favorable properties. Fabrication, testing, and operation are governed by strict procedures.

13.3   CONSUMPTION OF NUCLEAR FUELS

The generation of energy from nuclear fuels is unique in that a rather large amount of fuel must be present at all times for the chain reaction to continue. (In contrast, an automobile will operate even though its gasoline tank is practically empty.) There is a subtle relationship between reactor fuel and other quantities such as consumption, power, neutron flux, criticality, and control.

The first and most important consideration is the energy production, which is directly related to fuel consumption. Let us simplify the situation by assuming that the only fuel consumed is U-235, and that the reactor operates continuously and steadily at a definite power level. Since each atom "burned," i.e., converted into either U-236 or fission products by neutron absorption, has an accompanying energy release, we can find the amount of fuel that must be consumed in a given period.

Let us examine the fuel usage in a PWR with initial enrichment in U-235 of 3%. Suppose that the thermal power is 3000 MW and the reactor operates for 1 yr. Using the convenient rule of thumb that 1.3 g of U-235 is burned for each megawatt-day of energy, the weight of fuel used is

$$(3000 \text{ MW})(365 \text{ days})(1.3 \text{ g/MW-day}) = 1.4 \times 10^6 \text{ g}.$$

Now each gram of U-235 at 3% enrichment costs around $24 (see Table 10.1). The cost of the fuel consumed is around 33 million dollars. Adding the expense of fuel fabrication and storage brings the total to around 45 million dollars. A typical efficiency of conversion of thermal energy to electrical energy is $\frac{1}{3}$, so the electrical power is 1000 MW. Over the year (8760 hr) the energy delivered to the customers is 8.76×10^9 kWh, and thus the fuel cost is, per kilowatt hour, $0.0051, 0.51¢ or 5.1 mills. These figures are rough because the effect of generation and consumption of fissile plutonium has been ignored.

13.4 REACTOR CONTROL

Since no fuel is added during the operating cycle of the order of a year, the amount to be burned must be installed at the beginning. First, the amount of uranium needed to achieve criticality is loaded into the reactor. If then the "excess" is added, it is clear that the reactor would be supercritical unless some compensating action were taken. In the PWR, the excess fuel reactivity is reduced by the inclusion of control rods and boron solution.

The reactor is brought to full power and operating temperature and pressure by means of rod position adjustments. Then, as the reactor operates and fuel begins to burn out, the concentration of boron is reduced. By the end of the cycle, the extra fuel is gone, all of the available control absorption has been removed, and the reactor is shut down for refueling. The trends in fuel and boron are shown in Fig. 13.2, neglecting the effects of certain fission product absorption and plutonium production. The graph represents a case in which the power is kept constant. The fuel content thus linearly decreases with time. Such operation characterizes a reactor that provides a "base load" in an electrical generating system that also includes fossil fuel plants and hydroelectric stations.

The power level in a reactor was shown in Chapter 12 to be proportional to neutron flux. However, in a reactor that experiences fuel consumption the flux must increase in time, since the power is proportional also to the fuel content.

The amount of control absorber required at the beginning of the cycle is proportional to the amount of excess fuel added to permit burnup for power production. For example, if the fuel is expected to go from 3% to 1.5% U-235, an initial boron atom number density in the moderator is about 1.0×10^{-4} (in units of 10^{24}). For comparison, the number of water molecules per cubic centimeter is 0.0334. The boron content is usually expressed in parts per million (i.e., micrograms of an additive per gram of diluent). For our example, using 10.8 and 18.0 as the molecular weights of boron and water, there are $10^6(10^{-4})(10.8)/(0.0334)(18.0) = 1800$ ppm.

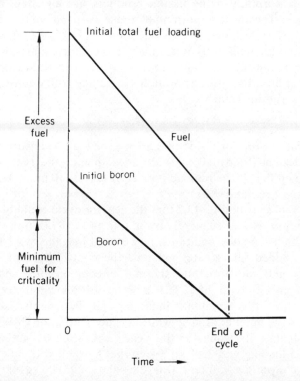

Fig. 13.2. Reactor control during fuel consumption in power reactor.

13.5 OTHER EFFECTS ON REACTOR OPERATION

The description of the reactor process just completed is somewhat idealized. Several other phenomena must be accounted for in design and operation.

If a reactor is fueled with natural uranium or slightly enriched uranium, the generation of plutonium tends to extend the cycle time. The fissile Pu-239 helps maintain criticality and provides part of the power. Small amounts of higher plutonium isotopes are also formed: Pu-240, fissile Pu-241 (14.355 year half life), and Pu-242. These isotopes and those of elements farther up the periodic table are called transuranic materials or actinides. They are important as fuels, poisons, or nuclear wastes.

Neutron absorption in the fission products has an effect on control requirements. The most important of these is a radioactive isotope of xenon, Xe-135, which has a cross section at 0.0253 eV of around 2.6 *million* barns. Its yield in fission is high, $y = 0.06$, meaning that for each fission, one obtains 6% as many atoms of Xe-135. In steady operation at high neutron flux, its rate of production is equal to its consumption by neutron absorption. Hence

$$N_X \sigma_{aX} = N_F \sigma_{fF} y.$$

Using the ratio σ_f / σ_a for U-235 of 0.86, we see that the absorption rate of Xe-135 is around $(0.86)(0.06) = 0.05$ times that of the fuel itself. This factor is about 0.04 if the radioactive decay ($t_H = 9.10$ hr) of xenon-135 is included (see Problem 13.3).

It might appear from Fig. 13.2 that the reactor cycle could be increased to as long a time as desired merely by adding more U-235 at the beginning. There are limits to such additions, however. First, the more the excess fuel that is added, the greater must be the control by rods or soluble poison. Second, radiation and thermal effects on fuel and cladding materials increase with life. The amount of allowable total energy extracted from the uranium, including all fissionable isotopes, is expressed as the number of megawatt-days per metric ton (MWd/tonne).†
We can calculate its value for the year's operation by noting that the initial U-235 loading was 2800 kg, twice that burned; with an enrichment of 0.03, the *uranium* content was 2800/0.03 = 93,000 kg or 93 tonnes. Using the energy yield of (3000 MW)(365 days) $\cong$ 1,100,000 MWd, we find 12,000 MWd/tonne. Taking account of plutonium and the management of fuel in the core, a typical average exposure is actually 30,000 MWd/tonne. It is desirable to seek larger values of this quantity, in order to prolong the cycle and thus minimize the cost of fuel reprocessing and fabrication.

13.6 THE NATURAL REACTOR

Until recently, it had been assumed that the first nuclear reactor was put into operation by Enrico Fermi and his associates in 1942. It appears, however, that a natural chain reaction involving neutrons and uranium took place in the African state of Gabon, near Oklo, some 2 billion years ago. At that time, the isotope concentration of U-235 in natural uranium

†The metric ton (tonne) is 1000 kg.

was higher than it is now because of the differences in half lives: U-235, 7.04×10^8 years; U-238, 4.47×10^9 years. The water content in a rich vein of ore was sufficient to moderate neutrons to thermal energy. It is believed that this "reactor" operated off and on for thousands of years at power levels of the order of kilowatts. The discovery of the Oklo phenomenon resulted from the observations of an unusually low U-235 enrichment in the mined uranium. The effect was confirmed by the presence of fission products.

13.7 SUMMARY

Reactors are classified according to their important features such as purpose, neutron energy, moderator and coolant, fuel, arrangement, and structural material. The principal types are the pressurized water reactor, the boiling water reactor, the high-temperature gas-cooled reactor, and the liquid metal fast breeder reactor. Excess fuel is added to a reactor initially to take care of burning during the operating cycle, with adjustable control absorbers present to maintain criticality. Account must be taken of fission product absorbers, especially the high cross section Xe-135, and of limitations on the energy release set by thermal and radiation effects. About 2 billion years ago deposits of uranium in Africa had high enough enrichment in U-235 to become natural chain reactors.

13.8 PROBLEMS

13.1. How many individual fuel pellets are there in the PWR reactor described in the text? Assuming a density of uranium oxide of 10 g/cm^3, estimate the total mass of uranium and U-235 in the core in kilograms. What is the initial fuel cost? (see Chapter 10).

13.2. How much money would be saved each year in producing the same electrical power: (a) if the thermal efficiency of a reactor could be increased from $\frac{1}{3}$ to 0.4? (b) if the fuel consumed were 2% enrichment rather than 3%?

13.3. (a) Taking account of Xe-135 production, absorption, *and decay*, show that the balance equation is

$$N_x(\phi\sigma_{ax} + \lambda_x) = \phi N_F \sigma_{fF} y.$$

(b) Calculate λ_x and the ratio of absorption rates in Xe-135 and fuel if ϕ is 2×10^{13} cm^{-2}-sec^{-1}.

13.4. Soon after a reactor starts operating, the fission product Xe-135 builds up to its essentially steady level. What amount of reduction in the boron content must be made to compensate for this new absorption?

13.5. The initial concentration of boron in a 10,000 ft^3 reactor coolant system is 1500 ppm (the number of micrograms of additive per gram of diluent). What volume of solution of concentration 8000 ppm should be added to achieve a new value of 1600 ppm?

13.6. An adjustment of boron content from 1500 to 1400 ppm is made in the reactor described in Problem 13.5. Pure water is pumped in and then mixed coolant and poison are pumped out in two separate steps. For how long should the 500 ft^3/min pump operate in each of the operations?

13.7. Find the ratio of weight percentages of U-235 and U-238 at a time 1.9 billion years ago, assuming the present 0.711/99.3.

14

Energy Conversion Methods

Most of the energy released in fission or fusion appears as kinetic energy of a few high-speed particles. As these pass through matter, they slow down by multiple collisions and impart thermal energy to the medium. It is the purpose of this chapter to discuss the means by which this energy is transferred to a cooling agent and transported to devices that convert mechanical energy into electrical energy.

14.1 METHODS OF HEAT TRANSMISSION

We learned in basic science that heat, as one form of energy, is transmitted by three methods—conduction, convection, and radiation. The physical processes for each of these are different: In *conduction*, molecular motion in a substance at a point where the temperature is high causes motion of adjacent molecules, and a flow of energy toward a region of low temperature takes place. The rate of flow is proportional to the slope of the temperature, i.e., the temperature gradient. In *convection*, molecules of a cooling agent such as air or water strike a heated surface, gain energy, and return to raise the temperature of the coolant. The rate of heat removal is proportional to the difference between the surface temperature and that of the surrounding medium, and also dependent on the amount of circulation of the coolant in the vicinity of the surface. In *radiation*, molecules of a heated body emit and receive electromagnetic radiations, with a net transfer of energy that depends on the temperatures of the body and the adjacent regions, specifically on the difference between the temperatures raised to the fourth power. For reactors, this mode of heat transfer is generally of less importance than the other two.

14.2 CONDUCTION IN REACTOR FUEL

The transfer of heat by conduction in a flat plate (insulated on its edges) is reviewed. If the plate has a thickness x and cross-sectional area A, and the temperature difference between its faces is ΔT, the rate of heat flow through the plate, Q, is given by the relation

$$Q = kA \frac{\Delta T}{x},$$

where k is the conductivity, with typical units joules/sec-°C-cm. For the plate, the slope of the temperature is the same everywhere. In a more general case, the slope may vary with position, and the rate of heat flow per unit area Q/A is proportional to the slope or gradient written as $\Delta T/\Delta x$.

The conductivity k varies somewhat with temperature but for our estimates it is assumed to be constant.

This idea is applied to the conduction in a single fuel rod of a reactor, with the rate of supply of thermal energy by fission taken to be uniform throughout the rod. If the rod is long in comparison with its radius R, or if it is composed of a stack of pellets, most of the heat flow is in the radial direction. If the surface is maintained at a temperature T_s by the flow of coolant, the center of the rod must be at some higher temperature T_0. As expected, the temperature difference is large if the rate of heat generation per unit volume q or the rate of heat generation per unit length $q_1 = \pi R^2 q$ is large. We can show† that

$$T_0 - T_s = \frac{q_1}{4\pi k},$$

and that the temperature T is in the shape of a parabola within the rod. Figure 14.1 shows the temperature distribution.

Let us calculate the temperature difference $T_0 - T_s$ for a reactor fuel rod of radius 0.5 cm, at a point where the power density is $q = 200 \text{ W/cm}^3$. This corresponds to a linear heat rate $q_1 = \pi R^2 q = \pi(0.25)(200) = 157 \text{ W/cm}$ (or 4.8 kW/ft). Letting the conductivity of UO_2 be $k = 0.062 \text{ W/cm-°C}$, we find $T_0 - T_s = 200°\text{C}$ (or 360°F). If we wish to keep the temperature low along the center line of the fuel, to avoid structural changes or melting, the conductivity k should be high, the rod size small,

---

†The amount of energy supplied within a region of radius r must flow out across the boundary. For a unit length of rod with volume πr^2 and surface area $2\pi r$, the generation rate is $\pi r^2 q$, equal to the flow rate $[-k(dT/dr)]2\pi r$. Integrating from $r = 0$, where $T = T_0$, we have $T = T_0 - (qr^2/4k)$. At the surface $T_s = T_0 - (qR^2/4k)$.

or the reactor power level low. In a typical reactor there is a small gap between the fuel pin and the inside surface of the cladding. During operation, this gap contains gases, which are poor heat conductors and thus there will be a rather large temperature drop across the gap. A smaller drop will occur across the cladding which is thin and has a high thermal conductivity.

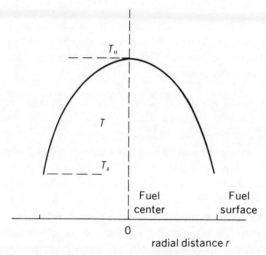

Fig. 14.1.   Temperature in fuel.

14.3 HEAT REMOVAL BY COOLANT

Convective cooling depends on many factors such as the fluid speed, the size and shape of the flow passage, and the thermal properties of the coolant, as well as on the area exposed and the temperature difference between surface and coolant $T_s - T_c$. Experimental measurements yield the "heat transfer coefficient" h, appearing in a working formula for the rate of heat transfer Q across a surface S,

$$Q = hS(T_s - T_c).$$

The units of h are typically W/cm^2-°C. In order to keep the surface temperature low, to avoid melting of the metal cladding of the fuel or to avoid boiling if the coolant is a liquid, a large surface area is needed or the heat transfer coefficient must be large, a low-viscosity coolant of good thermal conductivity is required, and the flow speed must be high.

As coolant flows along the many channels surrounding fuel pins in a reactor, it absorbs thermal energy and rises in temperature. Since it is the reactor power that is being extracted, we may apply the principle of conservation of energy. If the coolant of specific heat c enters the reactor at temperature T_c (in) and leaves at T_c (out), with a mass flow rate M, then the reactor thermal power P is

$$P = cM[T_c(\text{out}) - T_c(\text{in})] = cM\,\Delta T.$$

For example, let us find the amount of circulating water flow required to cool a reactor that produces 3000 MW of thermal energy. If the water enters at 300°C and leaves at 325°C, and the specific heat of water is 4185 J/kg-°C, the mass flow rate is

$$M = \frac{P}{c\,\Delta T} = \frac{3000 \times 10^6 \text{ W}}{\left(4185 \dfrac{\text{W-sec}}{\text{kg-°C}}\right)(25°C)} = 29,000 \frac{\text{kg}}{\text{sec}}.$$

Noting that 1 liter has a mass of 1 kg, this amounts to 1,740,000 liters/min. To appreciate the magnitude of this flow, we can compare it with that from a garden hose of 40 liters/min. The water for cooling a reactor is not wasted, of course, because it is circulated in a closed loop.

The temperature of coolant as it moves along any channel of the reactor can also be found by application of the above relation. In general, the power produced per unit length of fuel rod varies with position in the reactor because of the variation in neutron flux shape. For the special case of a *uniform* power along the z-axis with origin at the bottom (see Fig. 14.2a), the power per unit length is $P_1 = P/H$, where H is the length of fuel rod. The temperature rise of coolant at z with channel mass flow rate M is then

$$T_c(z) = T_c(\text{in}) + \frac{P_1 z}{cM},$$

which shows that the temperature increases linearly with distance along the channel (see Fig. 14.2b). The temperature difference between coolant and fuel surface is the same at all points along the channel for this power distribution, and the temperature difference between the fuel center and fuel surface is also uniform. We can plot these as in Fig. 14.2c. The highest temperatures in this case are at the end of the reactor.

If instead, the axial power were shaped as a sine function (see Fig. 14.3a) with $P \sim \sin(\pi z/H)$, the application of the relations for conduction and convection yield temperature curves as sketched in Fig. 14.3b. For this case, the highest temperatures of fuel surface and fuel center occur

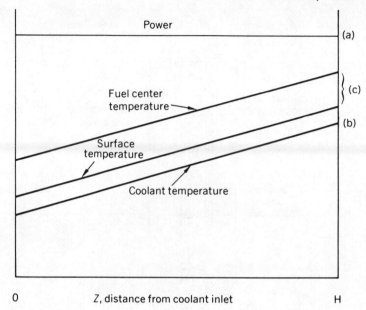

Fig. 14.2. Temperature distributions along axis of reactor with uniform power.

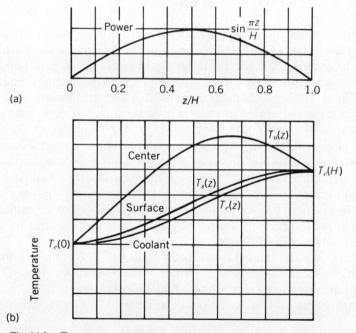

Fig. 14.3. Temperature distributions along channel with sine function power.

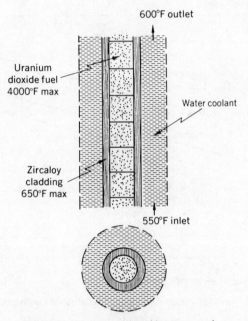

Fig. 14.4.   Reactor channel heat removal.

between the halfway point and the end of the reactor. In the design of a reactor, a great deal of attention is given to the determination of which channels have the highest coolant temperature and at which points on the fuel pins "hot spots" occur. Ultimately, the power of the reactor is limited by conditions at these channels and points. The mechanism of heat transfer from metal surfaces to water is quite sensitive to the temperature difference. As the latter increases, ordinary convection gives way to *nucleate boiling*, in which bubbles form at points on the surface, and eventually *film boiling* can occur, in which a blanket of vapor reduces heat transfer and permits hazardous melting. A parameter called "departure from nucleate boiling ratio" (DNBR) is used to indicate how close the heat flux is to the critical value. For example, a DNBR of 1.3 implies a safety margin of 30 percent. Figure 14.4 indicates maximum temperature values for a typical PWR reactor.

To achieve a water temperature of 600°F (about 315°C) requires that a very high pressure be applied to the water coolant-moderator. Figure 14.5 shows the behavior of water in the liquid and vapor phases. The curve that separates the two-phase regions describes what are called saturated conditions. Suppose that the pressure vessel of the reactor contains water at 2000 psi and 600°F and the temperature is raised to 650°F. The result will be considerable steam formation (flashing) within the liquid. The two-phase condition could lead to inadequate cooling of the reactor fuel. If instead the pressure were allowed to drop, say to 1200 psi, the vapor region is again entered and flashing would occur. However, it should be noted that deliberate two-phase flow conditions are used in boiling water reactors, providing efficient and safe cooling.

14.4 STEAM GENERATION AND ELECTRICAL POWER PRODUCTION

Thermal energy in the circulating reactor coolant is transferred to a working fluid such as steam, by means of a heat exchanger or steam generator. In simplest construction, this device consists of a vessel partly

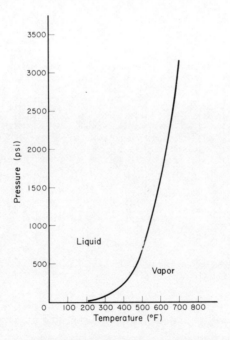

Fig. 14.5.   Relationship of pressure and temperature for water.

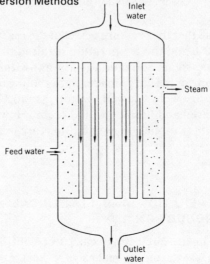

Fig. 14.6. Heat exchanger or steam generator.

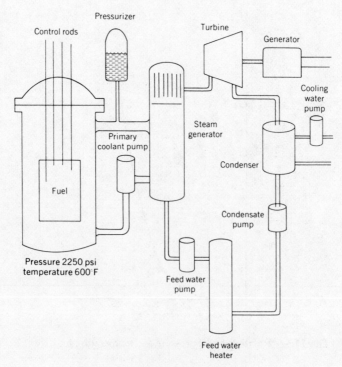

Fig. 14.7. PWR system flow diagram.

filled with water, through which many tubes containing heated water from the reactor pass, as in Fig. 14.6. Steam is evolved and flows to a turbine, while the water returns to the reactor. The conversion of thermal energy of steam into mechanical energy of rotation of a turbine and then to electrical energy from a generator is achieved by conventional means. Steam at high pressure and temperature is allowed to strike the blades of a turbine, which drives the generator. The exhaust steam is passed through a heat exchanger that serves as condenser, and the condensate is returned to the steam generator as feed water. Cooling water for the condenser is pumped from a nearby river, lake, or pond, which eventually receives the

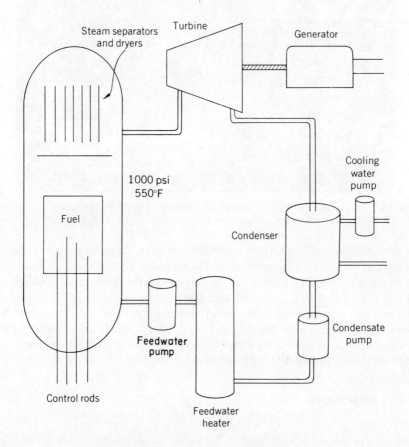

Fig. 14.8.   BWR system flow diagram.

Fig. 14.9. A nuclear power plant. (Courtesy of Yankee Atomic Electric Company and the United States Atomic Energy Commission.)

waste heat from the energy conversion process. In some installations, a special cooling tower is employed to discharge waste heat.

Figures 14.7 and 14.8 show the flow diagrams for the reactor systems of the PWR and BWR type. In the PWR, a pressurizer maintains the pressure in the system at the desired value. It uses a combination of immersion electric heaters and water spray system to control the pressure. Figure 14.9 shows the Yankee PWR nuclear power plant, at Rowe, Massachusetts, in operation since 1961.

14.5 SUMMARY

The principal modes by which fission energy is transferred in a reactor are conduction and convection. The radial temperature in a fuel pellet is approximately parabolic. The rate of heat transfer from fuel surface to coolant by convection is directly proportional to the temperature difference. The allowed power level of a reactor is governed by the temperatures at local "hot spots." Coolant flow along channels extracts thermal energy and delivers it to an external circuit consisting of a heat exchanger (PWR), a steam turbine that drives an electrical generator, a steam condenser, and various pumps.

14.6 PROBLEMS

14.1 Show that the temperature varies with radial distance in a fuel pin of radius R according to

$$T(r) = T_s + (T_0 - T_s)[1 - (r/R)^2],$$

where the center and surface temperatures are T_0 and T_s, respectively. Verify that the formula gives the correct results at $r = 0$ and $r = R$.

14.2. Explain the advantage of a circulating fuel reactor, in which fuel is dissolved in the coolant. What disadvantages are there?

14.3. If the power density of a uranium oxide fuel pin, of radius 0.6 cm, is 500 W/cm^3, what is the rate of energy transfer per centimeter across the fuel pin surface? If the temperatures of pin surface and coolant are 300°C and 250°C, what must the heat transfer coefficient h be?

14.4. A reactor operates at thermal power of 2500 MW, with water coolant mass flow rate of 15,000 kg/sec. If the coolant inlet temperature is 275°C, what is the outlet temperature?

14.5. A power reactor is operating with coolant temperature 500°F and pressure 1500 psi. A leak develops and the pressure falls to 500 psi. How much must the coolant temperature be reduced to avoid flashing?

15

Breeder Reactors

The most important feature of the fission process is, of course, the enormous energy release from each reaction. Another significant fact, however, is that for each neutron absorbed in a fuel such as U-235, more than two neutrons are released. In order to maintain the chain reaction, only one is needed. Any extra neutrons available can thus be used to produce other fissile materials such as Pu-239 and U-233 from the "fertile" materials, U-238 and Th-232, respectively. The nuclear reactions yielding the new isotopes were described in Chapter 7. If losses of neutrons can be reduced enough, the possibility exists for new fuel to be generated in quantities as large or even larger than the amount consumed, a situation called "breeding."

In this chapter we shall (a) examine the relationship between the reproduction factor and breeding, (b) describe the physical features of the liquid metal fast breeder reactor, and (c) look into the compatibility of uranium fuel resources and requirements.

15.1 THE CONCEPT OF BREEDING

The ability to convert significant quantities of fertile materials into useful fissile materials depends crucially on the magnitude of the reproduction factor, η, which is the number of neutrons produced per neutron absorbed in fuel. If ν neutrons are produced per fission, and the ratio of fission to absorption in fuel is σ_f / σ_a, then the number of neutrons per absorption is

$$\eta = \frac{\sigma_f}{\sigma_a} \nu.$$

The greater its excess above 2, the more likely is breeding. It is found that both ν and the ratio σ_f/σ_a increase with neutron energy and thus η is larger for fast reactors than for thermal reactors. Table 15.1 compares values of η for the main fissile isotopes in the two widely differing neutron energy ranges designated as thermal and fast. Inspection of the table shows that it is more difficult to achieve breeding with U-235 and Pu-239 in a thermal reactor, since the 0.08 or 0.12 neutrons are very likely to be lost by absorption in structural materials, moderator, and fission product poisons. A thermal reactor using U-233 is a good prospect, but the fast reactor using Pu-239 is the most promising candidate for breeding.

Table 15.1. Values of Reproduction Factor η.

	Neutron energy	
Isotope	Thermal	Fast
U-235	2.07	2.3
Pu-239	2.11	2.7
U-233	2.29	2.45

Absorption of neutrons in Pu-239 consists of both fission and capture, the latter resulting in the isotope Pu-240. If it captures a neutron, the fissile isotope Pu-241 is produced.

The ability to convert fertile isotopes into fissile isotopes can be measured by the *conversion ratio* (CR), which is defined as

$$CR = \frac{\text{fissile atoms produced}}{\text{fissile atoms consumed}}.$$

The fissile atoms are produced by absorption in fertile atoms; the consumption includes fission and capture. Examination of the neutron cycle for a thermal reactor (Fig. 12.4) shows that the conversion ratio is dependent on η_F, the reproduction factor of the *fissile* material used, on ϵ, the fast fission factor, and on the amount of neutron loss by leakage and by absorption in nonfuel material, the sum of which is represented by a term l. At the beginning of operation of a reactor, the conversion ratio is given by

$$CR = \eta_F\epsilon - 1 - l.$$

We can compare values of CR for different systems. For example, in a thermal reactor with $\eta_F = 2.07, \epsilon = 1.05$, and $l = 0.68$, the conversion ratio is 0.49. By adopting a reactor concept for which η_F is larger and by reducing neutron leakage and capture, a conversion ratio of 1 can be obtained, which means that a new fissile atom is produced for each one consumed. Further improvements yield a conversion ratio greater than 1. For example, if $\eta_F = 2.7$, $\epsilon = 1$, and $l = 0.3$, the conversion ratio would be 1.4, meaning that 40% more fuel is produced than is used. Thus the value of CR ranges from zero in the pure "burner" reactor, containing no fertile material, to numbers in the range 0.5–0.7 in typical "converter" reactors, to 1.0 or larger in the "breeder."

If unlimited supplies of uranium were available at very small cost, there would be no particular advantage in seeking to improve conversion ratios. One would merely burn out the U-235 in a thermal reactor, and discard the remaining U-238 or use it for non-nuclear purposes. Since the cost of uranium goes up as the accessible reserves decline, it is very desirable to use the U-238 atoms which comprise 99.28% of the natural uranium, as well as the U-235, which is only 0.72%. Similarly, the exploitation of thorium reserves is highly worthwhile.

We can gain an appreciation of how large the conversion ratio must be to achieve a significant improvement in the degree of utilization of uranium resources, by the following logic: Suppose that U-235 and Pu-239 are equally effective in multiplication in a thermal converter reactor, i.e., their η values are assumed to be comparable. Let 238 stand for the U-238 burned to produce fissile Pu-239 which is subsequently consumed, and let 235 stand for fissile U-235 consumed. Then the conversion ratio is

$$CR = \frac{238}{235 + 238} \text{ or } \frac{238}{235} = \frac{CR}{1 - CR}.$$

For example, if CR = 0.6

$$\frac{CR}{1 - CR} = \frac{0.6}{0.4} = 1.5.$$

This is far from complete conversion of the U-238. If all of the 0.72% U-235 in natural uranium feed were burned, the amount of U-238 converted would be only $(1.5)(0.72) = 1.08\%$, leaving about 99% unused. The percentage of the original *uranium* used is only $0.72 + 1.08 = 1.80$, or less than 2%. We can find what CR must be to achieve complete conversion. If all of the atoms in the 99.28% that is U-238 are converted while all of the atoms in the 0.72% that is U-235 are consumed it is necessary that

$$\frac{CR}{1-CR} = \frac{99.28}{0.72}.$$

Solving, CR = 0.9928, which is very close to unity. When one considers the effect of inevitable losses of uranium in reprocessing and refabrication, it is found that for practical purposes, the conversion ratio must be well above 1 in order to use all of the U-238.

When the conversion ratio is larger than 1, as in a fast breeder reactor, it is instead called the breeding ratio (BR), and the breeding gain (BG) = BR − 1 represents the extra plutonium produced per atom burned. The doubling time (DT) is the length of time required to accumulate a mass of plutonium equal to that in a reactor system, and thus provide fuel for a new breeder. The smaller the inventory of plutonium in the cycle and the larger the breeding gain, the quicker will doubling be accomplished. The technical term "specific inventory" is introduced, as the ratio of plutonium mass in the system to the electrical power output. Values of this quantity of 2.5 kg/MWe are sought. At the same time, a very long fuel exposure is desirable, e.g., 100,000 MWd/tonne, in order to reduce fuel fabrication costs. A breeding gain of 0.4 would be regarded as excellent, but a gain of only 0.2 would be very acceptable.

15.2 THE FAST BREEDER REACTOR

Several fast reactors have been operated successfully throughout the world. In the United States, the Experimental Breeder Reactor was the first power reactor to generate electricity; the Fermi reactor operated for some time; the Fast Flux Test Facility (FFTF) provides heat but no electricity. The Liquid Metal Fast Breeder Reactor (LMFBR) has been under design and construction for many years in a program entitled Clinch River Project located near Knoxville, Tenn. Other countries with fast power reactors are Great Britain, the Soviet Union, West Germany, Japan, and France.

The use of liquid sodium Na-23 as coolant ensures that there is little neutron moderation in the fast reactor. The element sodium melts at 208°F (98°C), boils at 1618°F (883°C), and has excellent heat transfer properties. With such a high melting point, pipes containing sodium must be thermally insulated and heated electrically to prevent freezing. The coolant becomes radioactive as the result of neutron absorption, producing the 15-hr Na-24. Great care must be taken to prevent contact between sodium and

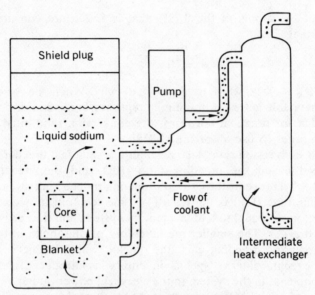

Fig. 15.1. Loop system for LMFBR.

water or air, which would result in a serious fire, accompanied by the spread of radioactivity. To avoid such an event, an intermediate heat exchanger is employed. Heat is transferred in it from radioactive sodium to nonradioactive sodium.

Two physical arrangements of the reactor core, pumps, and heat exchanger are possible, shown schematically in Figs. 15.1 and 15.2. The "loop" type is similar to the thermal reactor system, while in the "pot" type all of the components are immersed in a pool of liquid sodium. There are advantages and disadvantages to each concept, but both appear practical.

The French fast reactor Phenix at Marcoule is one of the most successful commercial breeder reactors. We shall describe its features in some detail. It is of the pot type, with a tank of 12.75 m diameter. The reactor has a core of height 85 cm and diameter 139 cm, fueled with a mixture of UO_2 and PuO_2 in the form of 0.55 cm diameter pellets encased in stainless steel tubes of wall thickness 0.045 cm. Bundles of 217 fuel pins form a fuel element, of which there are 103 total in the core. There are five control rods and three safety rods composed of B_4C. Above and below the core and surrounding it is the "blanket" containing depleted uranium in which neutron absorption occurs to produce new plutonium. The design power rating of Phenix is 250

MWe (563 MWt). The linear heat rate 430 W/cm is much higher than for a PWR (see Section 14.2). Similarly, since the reactor operates with fast neutrons, the flux is high, $6.7 \times 10^{15}/cm^2$-sec. The coolant temperature rises from 400°C to 560°C on passing through the reactor at a speed of 7.4 m/sec, flow rate 25 kg/sec. The energy is supplied to a turbine with steam at 510°C and pressure 163 kg/cm². The reactor is shut down every 6 months for refuelling. Plutonium produced in the blanket is extracted by reprocessing for use as new core fuel.

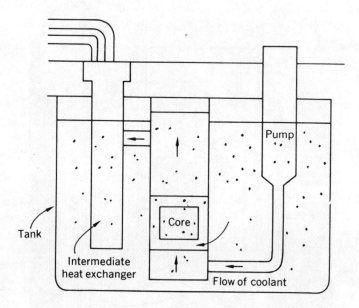

Fig. 15.2.   Pot system for LMFBR.

Although the principal attention throughout the world has been given to the liquid metal cooled fast breeder using U and Pu, other breeder concepts remain of interest. Included are (a) uranium and thorium fuel particles suspended in heavy water, (b) fuel and fertile elements as fluoride compounds mixed with other salts in molten form, and (c) a high-temperature gas-cooled graphite moderated reactor containing also a compound of beryllium, in which the (n, 2n) reaction occurs.

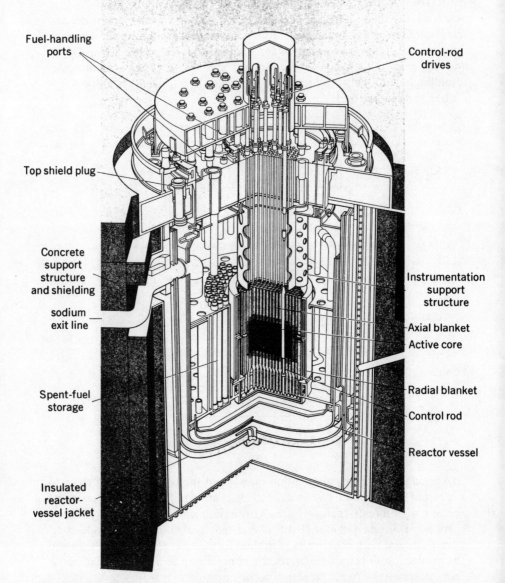

Fuel-handling ports

Control-rod drives

Top shield plug

Concrete support structure and shielding

sodium exit line

Spent-fuel storage

Insulated reactor-vessel jacket

Instrumentation support structure

Axial blanket

Active core

Radial blanket

Control rod

Reactor vessel

Fig. 15.3.   Liquid metal fast breeder reactor. (From *Fast Breeder Reactors* by Glenn T. Seaborg and Justin L. Bloom. Copyright © November 1970 by Scientific American, Inc. All rights reserved.)

15.3 BREEDING AND URANIUM RESOURCES

The urgency of establishing technical feasibility and commercial viability of the breeder can be seen by a comparison of uranium reserves and uranium requirements. Most of the uranium deposits in the United States are in sandstones of New Mexico and Wyoming, with lesser amounts in Texas, Utah, and Colorado. Estimates of uranium resources in the U.S. are made regularly by the Department of Energy. Counting proven reserves plus probable potential resources, the amount available at costs less than $15/lb U_3O_8 is 416,000 tons; less than $30 is 1,699,000 tons; less than $50, 2,425,000 tons. These figures must be compared with the future demand for natural uranium. Needs depend on the growth pattern, which is somewhat uncertain, leading to a range of estimates. If we accept the conservative estimate of nearly 200,000 MW(e) for the U.S. by the year 2000, the cumulative requirement is 800,000 tons of U_3O_8. We see that the price of fuel is likely to approach $30/lb in the U.S. by the year 2000.

On a world basis the situation is not as favorable. The "reasonably assured" world U resources for noncommunist countries are given in Table 15.2. The expected cumulative need is 2276 thousand tonnes, and thus the available 1650 thousand tonnes is only 73% of that required.

Table 15.2. World Uranium Resources†

Country	Thousands tonnes U_3O_8
United States	523
South Africa	306
Australia	289
Canada	167
Niger	160
France	37
India	30
Algeria	28
Gabon	20
Brazil	18
Argentina	18
Other	54
	1650

† From *Uranium 1978*, Atomic Industrial Forum, May 1979.

The cost of uranium is likely to become excessive by the turn of the century. It is generally agreed that the converter reactor of the LWR type is not suitable for use for much of the next century because of fuel resource limitations. The breeder reactor has the ability to use all of the uranium rather than a few percent. It could be used for the time required to develop alternatives sources such as fusion, solar energy, or geothermal energy. The impact of the breeder can be viewed in either of two ways. The demand for natural uranium would be reduced by a factor of 30, reducing the environmental effect of uranium mining while cutting down on fuel costs. Or, the span of time the relatively inexpensive fuel is available would be extended from some 25 years to perhaps 750 years. We note also that the U.S. has a large stockpile of depleted uranium, as tails from the gaseous diffusion isotope separation. Such material is as valuable as natural U in a breeder blanket.

It is clear that the importation of uranium by the U.S. is not practical since the U.S. already has about one-third of the resources. Energy needs of the rest of the world will easily absorb the rest of the fuel. The breeder reactor can fill the energy gap until longer-term sources such as solar or fusion energy can be developed. In the next chapter the prospects of nuclear fusion are considered.

15.4 SUMMARY

If the value of the neutron reproduction factor η is larger than 2 and losses of neutrons are minimized, breeding can be achieved in which more fuel is produced than is consumed. The conversion ratio (CR) measures the ability of a reactor system to transform a fertile isotope, e.g., U-238, into a fissile isotope, e.g., Pu-239. Complete conversion requires a value of CR of nearly 1. A fast breeder reactor using liquid sodium as coolant is being developed for commercial power. Estimated low-cost uranium reserves are sufficient only for a few decades of operation of reactors unless the breeder is successfully developed.

15.5 PROBLEMS

15.1. What are the largest conceivable values of the conversion ratio and the breeding gain?

15.2. An "advanced converter" reactor is proposed that will utilize 50% of the natural uranium supplied to it. Assuming all the U-235 is used, what must the conversion ratio be?

15.3. Explain why the use of a natural uranium "blanket" is an important feature of a breeder reactor.

15.4. Compute η and BG for a fast Pu-239 reactor if $\nu = 2.98$, $\sigma_f = 1.85$, $\sigma_c = 0.26$, and $l = 0.41$. (Note that the fast fission factor ϵ need not be included.)

15.5. With a breeding ratio BR = 1.20, how many kilograms of fuel will have to be burned in a fast breeder reactor operating only on plutonium in order to accumulate an extra 1260 kg of fissile material? If the power of the reactor is 1250 MWt, how long will it take in days and years, noting that it requires approximately 1.3 g of plutonium per MWd?

15.6. Verify the relation for conversion ratio CR = $\eta_F\epsilon - 1 - l$ by study of the thermal reactor neutron cycle (Chapter 12), noting that $l = l_f + l_t + a_t$, where the three terms on the right are the amounts of fast leakage, thermal leakage, and absorption in nonfuel materials per neutron absorbed in U-235. Note that $k = 1$ for a critical reactor.

15.7. If only U_3O_8 up to $30/lb cost were used, how much shortfall in tonnes of U_3O_8 is expected by the year 2000 in all of the world besides the U.S. (and communist countries?) Note the conversion 1 tonne = 1.1 tons.

16

Fusion Reactors

A device that permits the controlled release of fusion energy is designated as a fusion reactor, in contrast with one yielding fission energy, the fission reactor. As discussed in Chapter 8, the potentially available energy from the fusion process is enormous. The possibility of achieving controlled thermonuclear power on a practical basis has not yet been demonstrated, but progress in recent years gives encouragement that fusion reactors will be in operation early in the twenty-first century. In this chapter we shall review the choices of nuclear reaction, study the requirements for feasibility and practicality, and describe the physical features of machines that have been tested.

16.1 COMPARISON OF FUSION REACTIONS

The main nuclear reactions that combine light isotopes to release energy, as described in Chapter 8, are the D–D, D–T, and D–^{3}He. There are advantages and disadvantages of each. The reaction involving only deuterium uses an abundant natural fuel, available from water by isotope separation. However, the energy yield from the two equally likely reactions is low (4.03 and 3.27 MeV). Also the reaction rate as a function of particle energy is lower for the D–D case than for the D–T case, as shown in Fig. 16.1. The quantity $\overline{\sigma v}$, dependent on cross section and particle speed, is a more meaningful variable than the cross section alone.

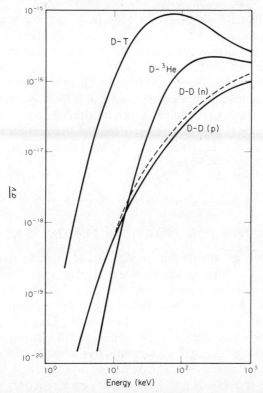

Fig. 16.1. Reaction rates for fusion reactions. The quantity $\overline{\sigma v}$, the average over a Maxwellian distribution of cross section times speed, when multiplied by particle densities gives the fusion rate per unit volume. (Adapted from *Fusion Cross Sections and Reactivities*, by George H. Miley, Harry Towner, and Nedad Ivich, Report C00-2218-17, 1974, University of Illinois.)

The D–T reaction yields a helium ion and a neutron with energies as shown

$$_1^2 H + _1^3 H \rightarrow {}_2^4 He + _0^1 n + 17.6 \, MeV$$
$$(3.5 \, MeV) \, (14 \, 1 \, MeV)$$

It has a very favorable energy yield and a high cross section, with a peak at around 100 keV. However, an artificial fuel, tritium, is required. The latter can be generated by neutron absorption in lithium, according to the two reactions

$$_3^6 Li + _0^1 n \rightarrow {}_1^3 H + _2^4 He + 4.8 \, MeV,$$

$$_3^7 Li + _0^1 n \rightarrow {}_1^3 H + _2^4 He + _0^1 n - 2.5 \, MeV.$$

We note that the neutron can come from the D–T reaction, suggesting a sort of breeding process. In the long run, use of the D–T reaction is limited by the availability of lithium resources, which though abundant are not nearly as inexhaustible as those of deuterium. All things considered, the D–T fusion reactor is the most likely to be operated first, and its success may lead to the development of a D–D reactor. Such a system is preferable since it requires only a natural isotope that is abundant in the world. There is little interest in the $D - {}^3He$ reation because of the need for a rare isotope as target and the low cross section, in spite of the high energy yield of 18.3 MeV.

16.2 REQUIREMENTS FOR PRACTICAL FUSION REACTORS

Since the purpose of any fusion device is to generate power, it is important to know the dependence of power density on factors such as plasma temperature and particle number densities. We first concentrate on conditions in the plasma, without reference to the surroundings. Our experience with the calculation of reaction rates (Chapter 5) may be applied. In a plasma containing n_D deuterons and n_T tritons per cubic centimeter, the particles interact with an effective average of the product of cross section and speed σv. The reaction rate per unit volume is $n_D n_T \sigma v$, and if the sensible energy yield is E for each reaction, the fusion power density is

$$p = n_D n_T \sigma v E.$$

We should assume here that the energy E is that of the helium atom only, since the neutron is free to escape from the plasma. On the basis of conservation of kinetic energy and momentum, the ${}^4_2He$ atom will have only $\frac{1}{5}$ of the reaction energy, i.e., $17.6/5 = 3.5$ MeV. When the plasma contains equal numbers of the heavy particles, $n_D = n_T = n$, then the power density p is proportional to n^2, the square of the particle number density.

The radiation losses described in Chapter 8 pose a serious limitation on the achievement of practical fusion power. The radiated power density p_r exceeds that from fusion, p, until the ignition temperature is reached. Ideally, we should like to maintain a steady electrical discharge with constant net power output. However, it will be sufficient if the reaction could be repeated periodically in short bursts of time of duration τ. If the frequency is high enough, the average power will be adequate. For a given value of τ and operating temperature T there must be a certain particle number density n for this mode of operation to be meaningful, according to the *Lawson criterion*

$$n\tau = \text{constant},$$

where the constant depends on T. The origin of this simple rule of thumb is as follows: We assume that energy must be provided to bring the plasma up to a temperature T where the fusion reaction is favorable. In one cubic centimeter there are $2n$ nuclei, each brought to average energy $\frac{3}{2}kT$, requiring an energy addition of $3nkT$. (We ignore electron energy in comparison with ion energy.) It is also necessary to supply energy to compensate for the radiation loss, as the product of power density p_r and time of pulse τ. Thus the total supplied must be $3nkT + p_r\tau$. After a pulse, we have the sum of the thermal energy of the plasma, the fusion energy, and the radiation energy, but with an efficiency of recovery ϵ that is less than 1, the available energy is only $\epsilon(3nkT + p\tau + p_r\tau)$. Equating the energies and noting that each power is proportional to n^2, we can solve for the product $n\tau$ as a function of T. The break-even point for the D–T reaction is around $n\tau = 10^{14}$ sec/cm^3. However, a fusion reactor with no energy extraction is useless, and it is necessary for $n\tau$ to be at least 10 times higher. The goal of 10^{15} sec/cm^3 can be reached with many different combinations of n and τ, e.g., $n = 10^{15}$, $\tau = 1$ sec or $n = 10^{21}$, $\tau = 10^{-6}$ sec. Figure 16.2 shows the accomplishments of some of the machines being tested as points on a graph of $n\tau$ against T, and shows the range of conditions that must be reached for feasibility of controlled fusion and the more difficult goal of a practical fusion reactor.

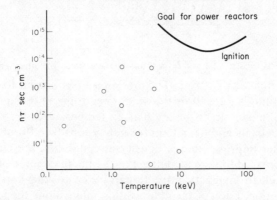

Fig. 16.2.   Progress of fusion toward a practical power reactor.

16.3 FUSION DEVICES

Many complex machines have been devised to generate a plasma and to provide the necessary electric and magnetic fields to achieve confinement of the discharge. We shall examine a few of these to illustrate the variety of possible approaches.

First, consider a simple discharge tube consisting of a gas-filled glass cylinder with two electrodes as in Fig. 16.3a. This is similar to the familiar fluorescent lightbulb. Electrons accelerated by the potential difference cause excitation and ionization of atoms. The ion density and temperature of the plasma that is established are many orders of magnitude below that needed for fusion. To reduce the tendency for charges to diffuse to the walls and be lost, a current-carrying coil can be wrapped around the tube, as sketched in Fig. 16.3b. This produces a magnetic field directed along the axis of the tube, and charges move in paths described by a helix, the

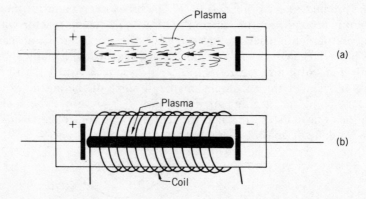

Fig. 16.3. Electrical discharges: (a) without magnetic field, (b) with magnetic field.

shape of a stretched coil spring. The motion is quite similar to that of ions in the cyclotron (Chapter 9) or the mass spectrograph (Chapter 10). The radii in typical magnetic fields and plasma temperatures are the order of 0.1 mm for electrons and near 1 cm for heavy ions (see Problem 16.1). In order to further improve charge density and stability, the current along the tube is increased to take advantage of the pinch effect, a phenomenon related to the electromagnetic attraction of two wires that carry current in the same direction. Each of the charges that move along the length of the tube constitutes a tiny current, and the mutual attractions provide a constriction in the discharge.

Neither of the above magnetic effects prevent charges from moving freely along the discharge tube, and losses of both ions and electrons are experienced at the ends. Two solutions of this problem have been tried. One is to wrap extra current-carrying coils around the tube near the ends (see Fig. 16.4a), increasing the magnetic field there. Figures 16.4b and

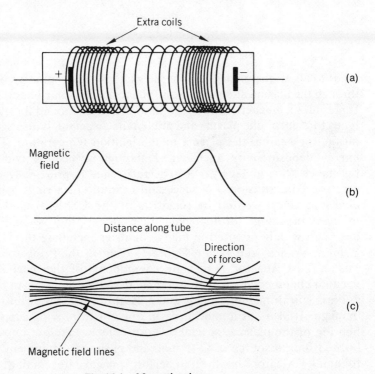

Fig. 16.4. Magnetic mirror.

16.4c show the field variation and the shape of the field lines. There is a tendency for charges to be forced back into the region of weak field, i.e., to be reflected. Such an arrangement is thus called a "mirror" machine, but it is by no means perfectly reflecting.

Another solution is to produce the discharge in a doughnut-shaped tube (torus), as shown in Fig. 16.5. Since the tube has no ends, the magnetic field produced by the coils is continuous, and the free motion of charges along the field lines does not result in any losses. Without any other forces acting, the variation in magnetic field in the radial direction in the tube

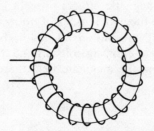

Fig. 16.5.   Plasma confinement in torus.

tends to allow charge diffusion to the walls. The introduction of a current through the plasma along the axis of the torus has some important effects. It provides a magnetic field in the form of circles around the current, that tends to "pinch" the plasma and prevent its dispersal. It also serves as the means for heating the plasma to the ignition temperature. The rate of energy deposition by a current I through a medium with electrical resistance R is given by I^2R. The resistance is principally due to the electron–ion collisions. A successful example of a ring-shaped fusion device is one invented by scientists of the USSR around 1960, the *tokamak* meaning "toroidal magnetic chamber" in Russian. The system has been widely adopted in many countries including the U.S., Great Britain, France, and Japan. Figure 16.6 shows the Princeton Tokamak Fusion Test Reactor. The main components of the machine are the vacuum chamber, magnet coils wrapped around the torus to produce an endless spiral magnetic field, and a shield to protect against radiation damage. It also has a neutral beam injector to provide supplementary heating of the plasma, special diverter coils to take out impurities and burned fuel, a source of D–T fuel, and electrical equipment. The MIT tokamak "Alcator" has the best performance to date, coming close to the breakeven energy condition.

The achievement of a self-sustaining fusion reaction in a plasma must be accompanied by the development of a practical method of extracting thermal energy. Let us visualize a system that will reveal some of the scientific and engineering problems to be solved. The reactor would probably consist of a long tube or circular ring, with many coaxial regions, as sketched in Fig. 16.7. In the highly ionized plasma, the D–T reaction would produce alpha particles and neutrons. The kinetic energies of the alpha particles would be transferred to the plasma and ultimately extracted. Neutrons would be slowed down and absorbed in a coolant such as liquid lithium that flows between two walls, extracting heat from the inner one,

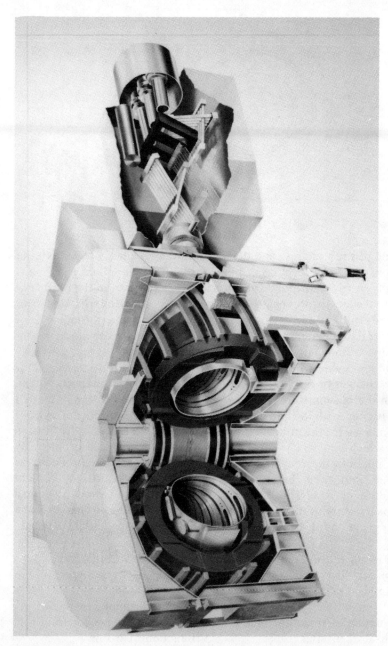

Fig. 16.6. Tokamak Fusion Test Reactor at Princeton University. (Courtesy Plasma Physics Laboratory.)

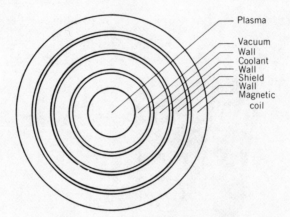

Plasma

Vacuum
Wall
Coolant
Wall
Shield
Wall
Magnetic
coil

Fig. 16.7. Hypothetical fusion reactor.

absorbing neutron energy, and delivering the total energy to an external heat exchanger. The reaction of neutrons with lithium could give rise to the supply of tritium needed for the basic D–T reaction in the plasma. A shield is needed to prevent the magnetic coils from overheating from conducted heat and electromagnetic radiation from the plasma.

The extraction of charged particles from a fusion reactor would permit conversion of nuclear-thermal energy into electrical energy. Efficiency and economics would both be improved since there would be no need for a heat exchanger or turbine-generator. Figure 16.8 shows one such system, in which the plasma from a mirror machine is allowed to expand into a conical region, with ion and electron collection electrodes in the walls.

A novel idea for the utilization of thermonuclear plasmas is the "fusion torch," which could help solve the world's waste-disposal problem by making possible the recycling of products. Waste materials would be converted into their basic elements through exposure to a high-temperature plasma. The torch might also perform various chemical reactions by applying ultraviolet (UV) radiation from the discharge. Some of the applications of the fusion torch that have been suggested are listed.

Uses of Plasma Energy	*Uses of UV Radiation*
Ore processing	Desalination of sea water
Waste product treatment	Radiative heating
Alloy separation	Waste sterilization
Destruction of toxic chemicals	Cultivation of algae for food
Disposal of plastics	Production of ozone

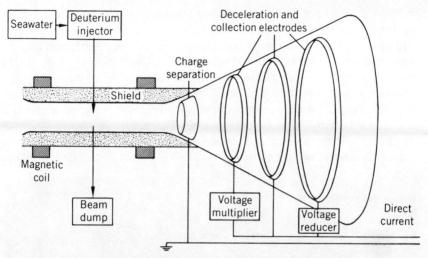

Fig. 16.8. Direct conversion fusion reactor. (From *The Prospects of Fusion Power* by William C. Gough and Bernard J. Eastlund. Copyright © February 1971 by Scientific American, Inc. All rights reserved.)

Even if net energy is produced in a fusion machine, there remain many scientific and engineering problems that must be solved before the energy system is commercially competitive. Among the areas of study facing the researchers are: material wear due to particle bombardment; protection of superconducting magnets; containment of radioactive tritium; removal of all impurities from the plasma; and cost reduction methods. The fusion reactor is said to be preferable to the fission reactor for several reasons — it has far less radioactivity; it is inherently safe against power rise; it has no decay heat; and it is not a source of weapons material. Unfortunately fusion reactors are not yet operable, and it is generally believed that they will only be commercially available after the year 2000.

16.4 LASER AND PARTICLE BEAM FUSION

Another approach to practical fusion is *inertial confinement*, in which very small pellets of a deuterium and tritium mixture are (as high-density gas or as ice) compressed and heated by laser light or by high-speed particles. The pellets act as miniature hydrogen bombs, exploding and delivering their energy to a wall and cooling medium. Figure 16.9 shows a quarter coin with some of the spheres. Their diameter is about 1/50 of a

millimeter (smaller than the periods on this page). To cause the thermonuclear reaction a beam of laser light is divided into many components
and directed simultaneously as a pulse of nanosecond† duration to
bombard a pellet from several directions. The laser energy is absorbed in
the outer layer of the sphere, causing some material to be evaporated
("ablated," as in the entry of a spacecraft into the earth's atmos-
sphere). As the particles escape from the surface they impart a reaction
momentum to the rest of the sphere. The resulting implosion force
compresses the material to an extremely high density, and the heating
ignites the center region of the pellet. The D–T nuclear reaction proceeds

Fig. 16.9. Gold microschells containing high-pressure D–T gas for use in laser fusion
(Courtesy of Los Alamos Scientific Laboratory.)

†1 nanosecond = 10^{-9} sec

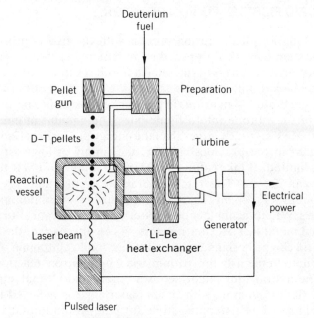

Fig. 16.10. Laser-fusion reactor.

outward to burn the rest of the fuel and to cause the microexplosion that lasts about 20 picoseconds.† The kinetic energy of the reaction products — neutrons and helium ions — is to be desposited in a layer of liquid lithium that is continuously circulated over the surface of the container and out to a heat exchanger. This isolation of the reaction from walls is expected to prevent material damage. Thus it may be necessary to replace wall metals frequently or to install exotic and expensive substances. Figure 16.10 shows a schematic arrangement of a laser-fusion reactor.

Research on laser fusion is being carried out at Lawrence Livermore Laboratory in California, using a machine called Shiva (after the many-armed Indian deity). Typical bombardments by 20 lasers onto a single pellet the size of a grain of sand deliver 26×10^{-12} watts for 95×10^{-12} seconds to obtain an energy deposition of 2500 joules and a resultant 7.5×10^9 fusion reactions. Research is also under way on inertial confinement using charged particles — electrons or ions — beams of which are available at the required energies. Problems include the proper focussing of the charge particle streams.

† 1 picosecond = 10^{-12} sec.

16.5 HYBRID FISSION–FUSION REACTOR

The combination of the fusion process with the fission process in a "hybrid" system has been proposed as an easier first step in achieving practical fusion. The idea is to surround a fusion reaction with a uranium or thorium blanket and allow the fast 14 MeV neutrons from the D–T reaction to generate 2–4 neutrons in the blanket, just as in a breeder, and convert a fertile isotope into a fissile one. The amount of energy per fission (190 MeV) is much larger than the energy per fusion (17.6 MeV). As was noted in the previous chapter, the supply of low-cost natural uranium is limited. It takes about 6000 tons to supply an LWR throughout its 30-year lifetime. To stretch the uranium supply, there are several choices: to reprocess spent fuel to separate the recycle plutonium, or to modify the LWR to achieve much better conversion ratios, or to construct liquid metal fast breeder reactors. As an alternative, the fission–fusion hybrid can be designed to produce new fissile plutonium or U-233 and as well to replenish the tritium used by neutron reactions with lithium. The hybrid thus might act as a "fuel factory" that supplies a number of fission reactors with fissile material. It is believed that the hybrid could serve as a stepping-stone for pure fusion in that valuable experience would be gained through its development.

16.6 SUMMARY

A fusion reactor, yet to be developed, would provide power from a controlled fusion reaction. Of the several possible nuclear reactions, the one that will probably be used first involves deuterium and tritium (produced by neutron absorption in lithium). A D–T reactor that yields significantly more energy than is supplied to achieve the reaction must have a product $n\tau$ above 10^{15}, where n is the particle number density and τ is the duration of pulsed operation. Many types of experimental machines have been tested. Most involve an electrical discharge (plasma) that is influenced by magnetic fields due to currents in external coils or within the discharge tube. A fusion reactor based on magnetic confinement of plasma would consist of a doughnut-shaped tube with several concentric layers. Research is also under way on inertial confinement, with laser beams causing the explosion of minature D–T pellets. A hybrid fission–fusion reactor may be developed in the near future.

16.7   PROBLEMS

16.1. Noting that the radius of motion R of a particle of charge q and mass m in a magnetic field B is $R = mv/qB$ and that the kinetic energy of rotation in the x-y plane is $\frac{1}{2}mv^2 = kT$, find the radii of motion of electrons and deuterons if B is 10 Wb/m^2 and kT is 100 keV.

16.2. Show that the effective nuclear reaction for a fusion reactor using deuterium, tritium, and lithium-6 is

$$^2_1H + ^6_3Li \rightarrow 2\,^4_2He + 22.4\,MeV.$$

16.3. Verify the statement that in the D–T reaction the ^4_2He particle will have $\frac{1}{5}$ of the energy.

16.4. (a) Assuming that in the D–D fusion reaction the fuel consumption is 0.151 g/MWd (Problem 8.3), find the energy release in J/kg. By how large a factor is the value larger or smaller than that for fission?

(b) If heavy water costs \$100/kg, what is the cost of deuterium per kilogram?

(c) Noting 1 kWh $= 3.6 \times 10^6$ J, find from (a) and (b) the energy cost in mills/kWh.

Part III Nuclear Energy and Man

The discovery of nuclear reactions that yield energy, radiation, and radioisotopes is generally regarded as one of the most significant scientific contributions of the twentieth century, in that it showed the possibility of enormous human benefit or of world destruction. It is thus understandable that nuclear energy should be a controversial subject. Many people have deplored its initial use for military purposes, while others regarded the action as necessary under the existing circumstances. Some believe that the discovery of nuclear energy should somehow have been avoided, while others hold that the revelation of natural phenomena is inevitable. Many uninformed persons see no distinction between nuclear weapons and nuclear reactors, while others recognize that the two are entirely different. A small group of scientists would abandon the use of nuclear energy on the basis of risks, while many other knowledgeable persons believe that it should be applied to national and world energy needs.

The variety of viewpoints on nuclear energy is but a part of a larger picture—the recent growth in concern about the effects of science and technology, which are claimed by some to be the source of many of the problems of advanced countries. Such a reaction is natural when one learns the extent of waste release and the effect of environmental pollution on wildlife and human beings. There is no doubt that there exists a sequence of events starting with a scientific discovery, continuing with a technological and commercial application, and too often ending with a new environmental or social problem. It does not follow, however, that the investigations should not have been initiated, but rather that adequate information about possible side effects should have been developed and that positive recommendations to prevent harmful releases should have been made. Nor does it suggest that new beneficial technology should be discouraged, but that realistic appraisals of the costs of ensuring adequate protection be developed and that the public be made aware of the additional expense. Finally, excessive concern about the effect of the industrial byproducts is paradoxical in a world in which two-thirds of the population receives none of the benefits of health, freedom from drudgery, and high standard of living that we enjoy as fruits of a technological civilization.

Decisions as to the uses of science are subject to ethical and moral criteria, but science itself, as a process of investigation and a body of information that is developed, must be regarded as neutral. Every natural resource has mixed good and evil. For example, fire is most necessary and welcome for warmth of our homes and buildings but can devastate our forests. Water is required for survival of every living being but in the form of a flood can ruin our cities and land. Drugs can help cure diseases but can incapacitate or kill us. Explosives are valuable for mining and construction but are also a tool of warfare. So it is with nuclear energy—on one hand, we have the benefits of heat and radiation for many human needs; on the other, the possibility of bombs and radioactive fallout. The key to application for benefit or detriment lies in man's decisions, and the fear of evil uses should not deter us from taking full advantage of good uses.

In Part III we shall review the history of nuclear energy, examine its hazards and the means available for protection, and describe some of the many peaceful applications of nuclear energy to the betterment of mankind. Finally, we shall discuss the role of nuclear energy in the long-term survival of our species.

17

The History of Nuclear Energy

The development of nuclear energy exemplifies the consequences of scientific study, technological effort, and commercial application. We shall review the history for its relation to our cultural background, which should include man's endeavors in the broadest sense. The author subscribes to the traditional conviction that history is relevant. Present understanding is grounded in recorded experience, and while we cannot undo errors, we can avoid them in the future. We can hopefully establish concepts and principles about human attitudes and capability that are independent of time, to help guide future action. Finally, we can draw confidence and inspiration from the knowledge of what man has been able to accomplish.

17.1 THE RISE OF NUCLEAR PHYSICS

The science on which practical nuclear energy is based can be categorized as classical, evolving from studies in chemistry and physics for the last several centuries, and modern, that related to investigations over the last hundred years into the structure of the atom and nucleus. The modern era begins in 1879 with Crookes' achievement of ionization of a gas by an electric discharge. Thomson in 1897 identified the electron as the charged particle responsible for electricity. Roentgen in 1895 had discovered penetrating X-rays from a discharge tube, and Becquerel in 1896 found similar rays—now known as gamma rays—from an entirely different source, the element uranium, which exhibited the phenomenon of natural radioactivity. The Curies in 1898 isolated the radioactive element radium. As a part of his revolutionary theory of motion, Einstein in 1905 concluded that the mass of any object increased with its speed, and stated his now-famous formula $E = mc^2$, which expresses the equivalence of mass and energy. At that time, no experimental verification was available, and Einstein could not have foreseen the implications of his relation.

In the first third of the twentieth century, a host of experiments with the various particles coming from radioactive materials led to a rather clear understanding of the structure of the atom and its nucleus. It was learned from the work of Bohr and Rutherford that the electrically neutral atom is constructed from negative charge in the form of electrons surrounding a central positive nucleus, which contains most of the matter of the atom. Through further work by Rutherford in England around 1919, it was revealed that even though the nucleus is composed of particles bound together by forces of great strength, nuclear transmutations can be induced, e.g., the bombardment of nitrogen by helium yields oxygen and hydrogen.

In 1930, Bothe and Becker bombarded beryllium with alpha particles from polonium and found what they thought were gamma rays but which Chadwick in 1932 showed were neutrons. A similar reaction is now employed in a nuclear reactor to provide a source of neutrons. Artificial radioactivity was first reported in 1934 by Curie and Joliot. Alpha particles injected into nuclei of boron, magnesium, and aluminum gave new radioactive isotopes of several elements. The development of machines to accelerate charged particles to high speeds opened up new opportunities to study nuclear reactions. The cyclotron, developed in 1932 by Lawrence, was the first of a sequence of devices of ever-increasing capability.

17.2 THE DISCOVERY OF FISSION

During the 1930s, Enrico Fermi and his co-workers in Italy performed a number of experiments with the newly discovered neutron. He reasoned correctly that the lack of charge on the neutron would make it particularly effective in penetrating a nucleus. Among his observations was the neutron capture reaction and the efficiency for producing a great variety of radioisotopes by neutrons slowed in paraffin or water. Breit and Wigner provided the theoretical explanation of slow neutron processes in 1936. Fermi made measurements of the distribution of both fast and thermal neutrons and explained the behavior in terms of elastic scattering, chemical binding effects, and thermal motion in target molecules. During this period, many cross sections for neutron reactions were measured, including that of uranium, but the fission process was not identified.

It was not until January 1939 that Hahn and Strassmann of Germany reported that they had found the element barium as a product of neutron bombardment of uranium. Frisch and Meitner made the guess that fission was responsible for the appearance of an element that is only half as heavy as uranium. Fermi then suggested that neutrons might be emitted during the process, and the idea was born that a chain reaction that released great amounts of energy might be possible. The press picked up the idea, and many sensational articles were written. The information on fission, brought to the United States by Bohr on a visit from Denmark, prompted a flurry of activity at several universities, and by 1940 nearly a hundred papers had appeared in the technical literature. All of the qualitative characteristics of the chain reaction were soon learned—the moderation of neutrons by light elements, thermal and resonance capture, the existence of fission in U-235 by thermal neutrons, the large energy of fission fragments, the release of neutrons, and the possibility of producing transuranic elements, those beyond uranium in the periodic table.

17.3 THE DEVELOPMENT OF NUCLEAR WEAPONS

The discovery of fission, with the possibility of a chain reaction of explosive violence, was of especial importance at this particular time in history, since World War II had begun in 1939. Because of the military potential of the fission process, a voluntary censorship of publication on the subject was established by scientists in 1940. The studies that showed U-235 to be fissile suggested that the new element plutonium, discovered in 1941 by Seaborg, might also be fissile and thus also serve as a weapon material. As early as July 1939, four leading scientists—Szilard, Wigner, Sachs, and Einstein—had initiated a contact with President Roosevelt, explaining the possibility of an atomic bomb based on uranium. As a consequence a small grant of $6000 was made by the military to procure materials for experimental test of the chain reaction. (Before the end of World War II, a total of $2 billion had been spent, an almost inconceivable sum in those times.) After a series of studies, reports, and policy decisions, a major effort was mounted through the U.S. Army Corps of Engineers under General Groves. The code name "Manhattan District" (or "Project") was devised, with military security on all information.

Although a great deal was known about the individual nuclear reactions, there was great uncertainty as to the practical behavior: could a chain reaction be achieved at all? If so, could Pu-239 in adequate quantities be produced? Could a nuclear explosion be made to occur? Could U-235 be separated on a large scale? Attacks on these questions were initiated at several institutions, and design of production plants began almost concurrently, with great impetus provided by the involvement of the United States in World War II after the attack on Pearl Harbor in December 1941 by the Japanese. The distinct possibility that Germany was actively engaged in the development of an atomic weapon served as a strong stimulus to the work of American scientists, most of whom were in universities. They and their students dropped their normal work to enlist in some phase of the project.

The whole Manhattan Project consisted of parallel endeavors, with major effort in the United States and cooperation with the United Kingdom, Canada, and France. At the University of Chicago, tests preliminary to the construction of the first atomic pile were made; and on December 2, 1942, Fermi and his associates achieved the first chain reaction under the stands of Stagg Field. By 1944, the plutonium production reactors at Hanford, Washington had been put into operation, providing the new element in kilogram quantities. At the University of California at Berkeley, the electromagnetic separation "calutron" process for isolating U-235 was perfected, and government production plants at Oak Ridge, Tennessee were built in 1943. At Columbia University, the gaseous diffusion process for isotope separation was studied, forming the basis for the present production system, the first units of which were built at Oak Ridge. At Los Alamos, New Mexico, theory and experiment led to the development of the nuclear weapons, first tested at Alamogordo, New Mexico, on July 16, 1945, and later used at Hiroshima and Nagasaki in Japan.

The brevity of this account fails to reveal the dedication of scientists, engineers, and other workers to the accomplishment of national objectives, or the magnitude of the design and construction effort by American industry. Two questions are inevitably raised—Should the atom bomb have been developed? Should it have been used? Some of the scientists who worked on the Manhattan Project have expressed their feeling of guilt for having participated. Some insist that a lesser demonstration of the destructive power of the weapon should have been arranged, which would have been sufficient to end the conflict. Many others believed that the security of the United States was threatened and that the use of the weapon shortened World War II greatly and thus saved a large number of casualties on both sides. It is some comfort, albeit small, that the existence of nuclear weapons has served as a deterrent to a direct conflict between major powers for several decades.

The discovery of nuclear energy, with its tremendous potential for the betterment of mankind through new unlimited energy resources, and through radioisotopes and their radiation for research, medical diagnosis and treatment, and agricultural improvement, can very well have benefits that far outweigh the detriments, particularly if we have sense enough not to use nuclear weapons again.

17.4 PEACEFUL APPLICATIONS OF NUCLEAR ENERGY

One of the first important events in the United States after World War II ended was the creation of the United States Atomic Energy Commission. This civilian federal agency was charged with the management and development of the nation's nuclear programs on behalf of peaceful applications. Several national laboratories were established to continue research in nuclear energy, at sites such as Oak Ridge, Argonne (near Chicago), Los Alamos, and Brookhaven (on Long Island). One of the main objectives of the AEC was to attack the problem of producing practical commercial nuclear power through research and development. This work was started at national laboratories, with Oak Ridge first looking at a gas-cooled reactor. They later planned a high-flux reactor fueled by highly enriched uranium alloyed with and clad with aluminum, and using water as moderator and coolant. The reactor was eventually built in Idaho as the Materials Testing Reactor. Before then, however, the idea arose of using a similar system for powering a nuclear submarine, and through the determination of H. G. Rickover, then a Navy captain, a development project was pushed through at Argonne. It was found that zirconium was preferable to aluminum because of its compatibility with high-temperature water under pressure. During the short period 1948 to 1953, many technical problems were resolved and a prototype submarine reactor was built and tested in Idaho. The Westinghouse Electric Corporation assisted in the development, did the design and construction, and later produced the reactor for installation in the first nuclear submarine, the *Nautilus*, which went to sea in 1955. This pressurized water reactor (PWR) concept was adapted by Westinghouse for use as the first commercial power plant at Shippingport, Pennsylvania, beginning operation in 1957 at an electric power output of 60 MW. Uranium dioxide pellets as fuel were first introduced in this design.

In the decade of the 1950s several reactor concepts were tested and dropped for various reasons. One used an organic liquid diphenyl as a coolant on the basis of a high boiling point. Unfortunately, radiation caused deterioration of the compound. Another was the homogeneous aqueous reactor, with a uranium salt in water solution that was circulated through the core and heat exchanger. Deposits of uranium led to excess heating and corrosion of wall materials. The sodium–graphite reactor had liquid metal coolant and carbon moderator. Only one commercial reactor of this type was built. The high-temperature gas-cooled reactor, developed by General Atomic Company, has not been widely adopted, but is a promising alternative to light water reactors by virtue of its graphite moderator, helium coolant, and uranium–thorium fuel cycle.

Two other reactor research and development programs were under way at Argonne over the same period. The first program was aimed at achieving power plus breeding of plutonium, using the fast reactor concept with liquid sodium coolant. The first electric power from a nuclear source was produced in late 1951 in the Experimental Breeder Reactor, and the possibility of breeding was demonstrated. This work has served as the basis for the present fast breeder reactor development program. The second program consisted of an investigation of the possibility of allowing water in a reactor to boil and generate steam directly. The principal concern was with the fluctuations and instability associated with the boiling. Tests called BORAX were performed that showed that a boiling reactor could operate safely, and work proceeded that led to electrical generation in 1955. The General Electric Company then proceeded to develop the boiling water reactor (BWR) concept further, with the first commercial reactor of this type put into operation at Dresden, Illinois in 1960.

On the basis of the initial successes of the PWR and BWR, and with the application of commercial design and construction know-how, Westinghouse and General Electric were able, in the early 1960s, to advertise large-scale nuclear plants of power around 500 MWe that would be competitive with fossil fuel plants in the cost of electricity. Immediately thereafter, there was a rapid move on the part of the electric utilities to order nuclear plants, and the growth in the late 1960s was phenomenal. Orders for nuclear steam supply systems for the years 1965–1970 inclusive amounted to around 88 thousand MWe, which was more than a third of all orders, including fossil fueled plants. The corresponding nuclear electric capacity was around a quarter of the total United States capacity at the end of that period of rapid growth.

The period 1970–1980 saw considerable slowing in the rate of installation of nuclear plants in the United States. Reasons were (a) the very long time (~ 10 years) to design, license, and construct nuclear facilities; (b) the energy-conservation measures adopted as a result of the Arab oil embargo of 1973–74, resulting in a lower growth rate of demand for electricity; (c) escalating costs of nuclear plants relative to other types; and (d) public opposition in some areas. Estimates on the number of new plants to be built in the U.S. during the last two decades of the century vary from zero to several hundred.

The AEC served a useful national purpose over a 20-year period following World War II but experienced increasing criticism for its dual role as promoter and regulator of nuclear power. In 1975 it was split into two agencies — the Energy Research and Development Administration (ERDA) and the Nuclear Regulatory Commission (NRC). Subsequently, a Department of Energy (DOE) was created to encompass the management of all forms of energy, including nuclear, in 1977.

Space does not permit the description of the important international aspects of the development and use of nuclear energy. It should be mentioned, however, that the United Kingdom as early as the 1950s developed and built many large graphite reactors for electrical power while Canada led the investigation of heavy water moderated reactors. Both countries have continued to make important contributions to the technology. Several other European countries have strong nuclear programs, either independently, as in the case of France, or as a part of Euratom, a cooperative organization. The U.S.S.R. has a number of power reactors in use, and Japan is contributing strongly to the development of breeders. Valuable assistance to the nuclear field is provided by the International Atomic Energy Agency, based in Vienna.

Less than thirty years had elapsed between the discovery of the fission process and the advent of practical nuclear reactors for electrical power. The endeavor revealed a new concept—that large-scale national technological projects could be undertaken and successfully completed, by the application of large amounts of money and organization of the efforts of many sectors of society. The nuclear project in many ways served as a model for the United States space program of the 1960s. The most important lesson that the history of nuclear energy development may have for us is that urgent national and world problems can be solved by wisdom, dedication, and cooperation.

17.5   SUMMARY

A sequence of many investigations in atomic and nuclear physics spanning the period 1879–1939 led to the discovery of fission. New knowledge was developed about particles and rays, radioactivity, the equivalence of matter and energy, nuclear transmutation, and the structure of the atom and nucleus. The existence of fission suggested that a chain reaction involving neutrons was possible and that the process had military significance. A major national program was initiated that included uranium isotope separation by electromagnetic and gaseous diffusion, nuclear reactor studies, plutonium production, and weapons development, culminating in the use of the atomic bomb to end World War II. In the post-war period, emphasis was placed on peaceful applications of nuclear energy under the United States Atomic Energy Commission. Four reactor concepts—the pressurized water, boiling water, fast breeder, and gas cooled—evolved through work by national laboratories and industry. The first two were brought to commercial status in the 1960s. The nuclear power project of the United States demonstrated that major national goals could be achieved with sufficient effort. Many other countries also have active programs of nuclear power development.

18

Biological Effects of Radiation

All living species are exposed to a certain amount of natural radiation in the form of particles and rays. In addition to the sunlight, without which life would be impossible to sustain, all beings experience cosmic radiation from space outside the earth and natural background radiation from materials on the earth. There are rather large variations in the radiation from one place to another, depending on mineral content of the ground and on the elevation above sea level. Man and other species have survived and evolved within such an environment in spite of the fact that radiation has a damaging effect on biological tissue. The discovery by man of means to generate radiation, using X-ray machines, particle accelerators, or nuclear reactors, has added potential hazard to his existence. In assessing the importance of such man-made radiation, comparison is often made with levels in the naturally occurring background radiation.

We shall now describe the biological effect of radiation on cells, tissues, organs, and individuals, identify the units of measurement of radiation and its effect, and review the philosophy and practice of setting limits on exposure. Special attention will be given to regulations related to nuclear power plants.

A brief summary of modern biological information will be useful in understanding radiation effects. As we know, living beings consist of a great variety of species of plants and animals; they are all composed of cells, which carry on the processes necessary to survival. The simplest organisms such as algae and protozoa consist of only one cell, while complex beings such as man are composed of specialized organs and tissues that contain large numbers of cells, examples of which are nerve,

181

muscle, epithelial, blood, skeletal, and connective. The principal components of a cell are the *nucleus* as control center, the *cytoplasm* containing vital substances, and the surrounding *membrane,* as a porous cell wall. Within the nucleus are the *chromosomes,* which are long threads containing hereditary material. The growth process involves a form of cell multiplication called *mitosis*—in which the chromosomes separate in order to form two new cells identical to the original one. The reproduction process involves a cell division process called *meiosis*—in which germ cells are produced with only half the necessary complement of chromosomes, such that the union of sperm and egg creates a complete new entity. The laws of heredity are based on this process. The genes are the distinct regions on the chromosomes that are responsible for inheritance of certain body characteristics. They are constructed of a universal molecule called DNA, a very long spiral staircase structure, with the stairsteps consisting of paired molecules of four types. Duplication of cells in complete detail involves the splitting of the DNA molecule along its length, followed by the accumulation of the necessary materials from the cell to form two new ones. In the case of man, there are 46 chromosomes, containing about four billion of the DNA molecule steps, in an order that describes each unique person.

18.1 PHYSIOLOGICAL EFFECTS

The various ways that moving particles and rays interact with matter discussed in earlier chapters can be reexamined in terms of biological effect. Our emphasis previously was on what happened to the radiation. Now, we are interested in the effects on the medium, which are viewed as "damage" in the sense that disruption of the original structure takes place, usually by *ionization.* We saw that energetic electrons and photons are capable of removing electrons from an atom to create ions; that heavy charged particles slow down in matter by successive ionizing events; that fast neutrons in slowing impart energy to target nuclei, which in turn serve as ionizing agents; that the loss of gamma ray may be accompanied by an electron–positron pair as new radiation; and that capture of a slow neutron results in a gamma ray and a new nucleus that recoils with appreciable energy.

As a good rule of thumb, 32 eV of energy is required on the average to create an ion pair. This figure is rather independent of the type of ionizing radiation, its energy, and the medium through which it passes. For instance, a single 4-MeV alpha particle would release about 10^5 ion pairs before stopping. Part of the energy goes into molecular excitation and the formation of new chemicals. Water in cells can be converted into free radicals such as H, OH, H_2O_2, and HO_2. Since the human body is largely water, much of the effect of radiation can be attributed to the chemical reactions of such products. In addition, direct damage can occur, in which the radiation strikes certain molecules of the cells, especially the DNA that controls all growth and reproduction.

The most important point from the biological standpoint is that the bombarding particles have energy, which can be transferred to atoms and molecules of living cells, with a disruptive effect on their normal function. Since an organism is composed of very many cells, tissues, and organs, a disturbance of one atom is likely to be imperceptible, but exposure to many particles or rays can alter the function of a group of cells and thus affect the whole system. It is usually assumed that damage is cumulative, even though some accommodation and repair surely takes place.

The physiological effects of radiation may be classified as *somatic*, which refers to the body and its state of health, and *genetic*, involving the genes that transmit hereditary characteristics. The somatic effects range from temporary skin reddening when the body surface is irradiated, to a life shortening of an exposed individual due to general impairment of the body functions, to the initiation of cancer in the form of tumors in certain organs or as the blood disease, leukemia. The term "radiation sickness" is loosely applied to the immediate effects of exposure to very large amounts of radiation. The genetic effect consists of mutations, in which progeny are significantly different in some respect from their parents, usually in ways that tend to reduce the chance of survival. The effect may extend over many generations.

Although the amount of ionization produced by radiation of a certain energy is rather constant, the biological effect varies greatly with the type of tissue involved. For radiation of low penetrating power such as alpha particles, the outside skin can receive some exposure without serious hazard, but for radiation that penetrates tissue readily such as X-rays, gamma rays, and neutrons, the critical parts of the body are bone marrow as blood-forming tissue, the reproductive organs, and the lenses of the eyes. The thyroid gland is important because of its affinity for the fission product iodine, while the gastro-intestinal tract and lungs are sensitive to radiation from radioactive substances that enter the body through eating or breathing.

18.2 RADIATION DOSAGE UNITS

A number of specialized terms need to be defined in order to be able to discuss the biological effects of radiation. First is the absorbed *dose* (*D*). It is the amount of energy imparted to each gram of exposed biological tissue, and it appears as excitation or ionization of the molecules and atoms of the tissue. To illustrate, suppose that an organ weighing 0.05 kg was exposed to radiation from a radioactive material and there resulted in a release of 0.01 J of energy. In conventional terms, the dose would be 0.01/0.05=0.2 J/kg. A special unit that is convenient in dealing with energy absorption is the *rad*, which is 0.01 J/kg. For the example, the dose to the organ would thus be 20 rads.

The biological effect of energy deposition may be large or small depending on the type of radiation. For instance, a rad of dosage due to fast neutrons or alpha particles is much more damaging than a rad of dosage by X-rays or gamma·rays. In general, heavy particles create a more serious effect than do photons because of the greater energy loss with distance and resulting higher concentration of ionization. The "dose equivalent" (DE) is the product of the dose and a number that expresses the relative biological importance of the radiation. One of these is called a quality factor (QF) (see Table 18.1). The unit of measure of DE is the *rem* (an acronym for roentgen†-equivalent-man). For example, if QF were 3 for the radiation in the above example, the DE would be 60 rem. The millirem (1 mrem = 1/1000 rem) is a frequently used unit in describing small radiation doses.

Table 18.1.   Quality Factors.

X- and gamma rays	1
Beta particles > 30 keV	1
Beta particles < 30 keV	1.7
Thermal neutrons	3
Fast neutrons, protons, alpha particles	10
Heavy ions	20

The long-term effect of radiation on an organism also depends on the rate at which energy is deposited. Thus the *dose rate*, expressed in convenient units such as rads per hour or millirems per year, is used. Note that if dose is an energy, the dose rate is a power.

†The roentgen is a unit of exposure dose that was devised at a time when the principal radiations were X-rays and gamma rays. The rad and the rem are preferred units.

We shall describe the methods of calculating dosage in the following chapter. For perspective, however, we can cite some typical figures. A single sudden exposure that gives the whole body of a person 20 rem will give no perceptible clinical effect, but a dose of 400 rem will probably be fatal; natural radiation background provides about 100 mrem/yr; present medical and dental practice on the average gives nearly this same amount of additional dosage through the use of X-rays for diagnosis; regulations limit the dose rate above natural background to 5 mrem/yr at the site boundary of a power reactor.

The amounts of energy that result in biological damage are remarkably small. A gamma dose of 400 rem, which is very large in terms of biological hazard, corresponds to 4 J/kg, which would be insufficient to raise the temperature of a gram of water as much as 0.001°C. This fact shows that radiation affects the function of the cells by action on certain molecules, not by a general heating process.

18.3   ESTABLISHMENT OF LIMITS OF EXPOSURE

A typical bottle of aspirin will specify that no more than two tablets every four hours should be administered, implying that a larger or more frequent "dose" would be harmful. Such a limit is based on experience accumulated over the years with many patients. Although radiation has medical benefit only in certain treatment, the idea of the need for a *limit* is similar.

As we seek to clean up the environment by controlling emissions of waste products from industrial plants, cities, and farms, it is necessary to specify water or air concentrations of materials such as sulfur or carbon monoxide that are below the level of danger to living beings. Ideally, there would be zero contamination, but it is generally assumed that some releases are inevitable in an industrialized world. Again, limits based on knowledge of effects on living beings must be set.

For the establishment of limits on radiation exposure, agencies have been in existence for many years. Examples are the International Commission on Radiological Protection (ICRP), the National Council on Radiation Protection and Measurements (NCRP), and the Federal Radiation Council (FRC). Their general procedure is to study data on the effects of radiation and to arrive at practical limits that take account of both the risk and benefit of using nuclear equipment and processes.

There have been many studies of the effect of radiation on animals other than man, starting with early observations of genetic effects on fruit flies. Small mammals such as mice provide a great deal of data rapidly. Since controlled experiments on man are unacceptable, most of the available information on somatic effects comes from improper practices or accidents. Data are available, for example, on the incidence of sickness and death from exposure of workers who painted radium on luminous-dial watches or of doctors who used X-rays without proper precautions. The number of serious radiation exposures in the nuclear industry is too small to be of use on a statistical basis. The principal source of information is the comprehensive study of the victims of the atomic bomb explosions in Japan in 1945. The incidence of fatalities as a function of dose is plotted on a graph similar to Fig. 18.1a, where the data are seen to lie only in the high dosage range. In the range below 10 rads, there is no statistical indication of any increase in incidence of fatalities over the number in unexposed populations. The nature of the curve in the low dose range is unknown, and one could draw the curves labeled "unlikely" and "likely" as in Fig. 18.1b. In order to be conservative, i.e., to overestimate effects of radiation in the interests of providing protection, a linear extrapolation through zero is made, the "assumed" curve.

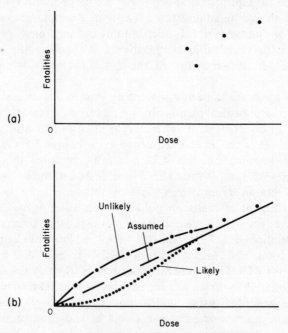

Fig. 18.1.   Radiation hazard analysis.

There is evidence that the biological effect of a given dose administered almost instantly is greater than if it were given over a long period of time. In other words, the hazard is less for low dose rates, presumably because the organism has the ability to recover or adjust to the radiation effects. If, for example (see Problem 18.2), the effect actually varied as the square of the dose, the linear curve would overestimate the effect by a factor of 100 in the vicinity of 1 rem. Although the hazard for low dose rates is small, and there is no clinical evidence of permanent injury, it is *not* assumed that there is a threshold dose, i.e., one below which no biological damage occurs. Instead, it is assumed that there is always some risk. The linear hypothesis is retained, in spite of the likelihood that it is overly conservative. There is a growing body of information on genetic effects in animals that tend to support this view.

The basic question then faced by standards-setting bodies is "what is the maximum acceptable upper limit for exposure?" One answer is zero, on the grounds that any radiation is deleterious. The view is taken that it is unwarranted to demand zero, as both maximum and minimum, because of the benefit from the use of radiation or from devices that have potential radiation as a byproduct.

The dose limits adopted for total body dose are 5000 mrem/yr for occupational workers, 500 mrem/yr to any individual member of the public, and an average over the whole population of 170 mrem/yr from all artificial sources other than medical applications. There are variations in permissible dose rate for workers according to the organ affected, as listed below.

Gonads, total body, and red bone marrow	5 rem/yr
Skin and bone	30 rem/yr
Other internal organs	15 rem/yr

The standards groups recommend that these figures be reduced by a factor of 10 for exposure of a person in the general public.

In the practical application of a dose rate limit standard such as the 500 and 170 mrem/yr, the question arises "what should be the limit set at the boundary of a nuclear plant?" It is clear that if the 500 mrem/yr individual figure is adopted, there would be a much lower level on the average throughout the country, since nuclear plants are widely distributed geographically, and exposure would certainly decrease greatly with distance from the site. The Nuclear Regulatory Commission specifies a considerably lower limit of 5 mrem/yr for the maximum dose rate at the site boundary of a nuclear power plant. Several comparisons can be made between this figure and other information. It is 1% of the individual limit or 3% of the whole population limit as recommended by ICRP and FRC. It is also about $\frac{1}{20}$ of the typical natural background. Since such low levels are not easily measured, it is necessary to calculate the dosage increase from the amount of radioactive material released. It is also comparable to the increase in dose received from cosmic radiation by a passenger on a single round trip jet airplane flight across the United States. The lower NRC standards are generally regarded as adequate for routine releases from nuclear power plants, even though zero release would be ideal. Estimates made by highly respected health physicists lead to the conclusion that the limits set by the NRC result in an exposure to the total population of the United States that statistically could result in about thirty additional deaths annually, in contrast to the millions of deaths annually due to the total of heart disease, cancer, stroke, and accidents. The effects of the slight extra exposure are believed to be completely masked by other hazards of existence.

18.4 SUMMARY

When radiation interacts with biological tissue, energy is deposited and ionization takes place that causes damage to cells. The effect on organisms is somatic, related to body health, and genetic, related to inherited characteristics. Radiation dose equivalent as a biologically effective energy deposition per gram is usually expressed in rem, with natural background giving about 0.1 rem/yr. Exposure limits are set by use of data on radiation effects at high dosages with a conservative linear hypothesis applied to predict effects at low dose rates.

18.5 PROBLEMS

18.1. A beam of 2-MeV alpha particles with current density 10^6 cm^{-2}-sec^{-1}, is stopped in a distance of 1 cm in air, number density 2.7×10^{19} cm^{-3}. How many ion pairs per cm^3 are formed? What fraction of the targets experience ionization?

18.2. If the chance of fatality from radiation dose is taken as 0.5 for 400 rem, by what factor would the chance at 2 rem be overestimated if the effect varied as the square of the dose rather than linearly?

18.3. A worker in a nuclear laboratory receives a whole-body exposure for 5 minutes by a thermal neutron beam at a rate 20 millirads per hour. What dose (in mrad) and dose equivalent (in mrem) does he receive? What fraction of the yearly dose limit of 500 mrem/yr for an individual is this?

18.4. A person receives the following exposures in millirems in a year: 1 medical X-ray, 100; drinking water 50; cosmic rays 30; radiation from house 60; K-40 and other isotopes 25; airplane flights 10. Find the percentage increase in exposure that would be experienced if he also lived at a reactor site boundary, assuming that the maximum NRC radiation level existed there.

19

Radiation Protection and Control

Protection of biological entities from hazard of radiation exposure is a fundamental requirement in the application of nuclear energy. Safety is provided by the use of one or more general methods that involve control of the source of radiation or its ability to affect living organisms. We shall identify these methods and describe the role of calculations in the field of radiation protection.

19.1 PROTECTIVE MEASURES

Radiation and radioactive materials are the link between a device or process as a source, and the living being to be protected from hazard. We can try to eliminate the source, or remove the individual, or insert some barrier between the two. Several means are thus available to help assure safety.

The first is to avoid the generation of radiation or isotopes that emit radiation. For example, the production of undesirable emitters from reactor operation can be minimized by the control of impurities in materials of construction and in the cooling agent. The second is to be sure that any radioactive substances are kept within containers or multiple barriers to prevent dispersal. Isotope sources and waste products are frequently sealed within one or more independent layers of metal or other impermeable substance, while nuclear reactors and chemical processing equipment are housed within leak-tight buildings. The third is to provide layers of shielding material between the source of radiation and the individual. The fourth is to restrict access to the region where the radiation level is hazardous, and take advantage of the reduction of intensity with distance. The fifth is to dilute a radioactive substance with very large volumes of air or water on release, to lower the concentration of harmful material. The sixth is to limit the time that a person remains within a radiation zone, to reduce the dose received. We thus see that radioactive materials may be treated in two quite different ways— *retention* and *dispersal*, while exposure to radiation can be avoided by methods involving *distance, shielding,* and *time.*

The analysis of radiation hazard and protection and the establishment of safe practices is part of the function of the science of radiological protection or health physics. Every user of radiation must follow accepted procedures, while health physicists provide specialized technical advice and monitor the user's methods. In the planning of research involving radiation or in the design and operation of a process, calculations must be made that relate the radiation source to the biological entity, using exposure limits provided by regulatory bodies. Included in the evaluation are the necessary protective measures for known sources, or limits that must be imposed on the radiation source, the rate of release of radioactive substances, or the concentration of radioisotopes in air, water, and other materials. Although the detailed calculations are very involved, a few simple examples will help us appreciate the approach used.

19.2 ENERGY DEPOSITION

The radiation dose or dose rate to tissue of a biological specimen is of central importance. We can find the rate of energy deposition in biological tissue by use of principles discussed in Chapter 5. Visualize a stream of radiation such as gamma rays as it passes through a substance. Flux and current are the same for this stream, both j and ϕ being the product nv. The reaction rate per unit volume is $\phi\Sigma$, where the cross section is formed from components that involve absorption of the photons. If the photon energy is E, the energy absorbed per cm³ per sec is $\phi\Sigma E$, per gram per sec is $\phi\Sigma E/\rho$ where ρ is the density of the material. Thus the dose in joules per gram on exposure for a time t is

$$D = \phi\Sigma Et/\rho.$$

This relation may be used to find dose, flux, or time.

For example, let us find the gamma ray flux that yields a limiting external dose of 170 mrem in 1 yr, with continuous exposure assumed. Suppose that the gamma rays have 1 MeV energy, and that the cross section for interaction with tissue of density $1.0\,\text{g/cm}^3$ is $0.03\,\text{cm}^{-1}$.

Letting the quality factor be 1 for this radiation the dose and dose equivalent are the same, 0.170 rem or 0.170 rad, i.e., $D = 1.7 \times 10^{-6}$ J/g. Also $E = 1$ MeV $= 1.60 \times 10^{-13}$ J. Solving.

$$\phi = \frac{D\rho}{\Sigma Et} = \frac{(1.7 \times 10^{-6} \text{ J/g}) (1 \text{ g/cm}^3)}{(0.03 \text{ cm}^{-1}) (1.60 \times 10^{-13} \text{J}) (3.16 \times 10^7 \text{ sec})}$$

or

$$\phi = 11.2 \text{ cm}^{-2}\text{- sec}^{-1}.$$

This value of the gamma ray flux may be scaled up or down if another dose limit is specified. The fluxes of various particles corresponding to 170 mrem/yr are shown in Table 19.1.

Table 19.1.   Radiation Fluxes (170 mrem/yr).

Radiation type	Flux (cm^{-2}- sec^{-1})
X- or gamma rays	11.2
Beta particles	0.25
Thermal neutrons	5.2
Fast neutrons	0.15
Alpha particles	1.2×10^{-5}

19.3   EFFECT OF DISTANCE

For protection, advantage can be taken of the fact that radiation intensities decrease with distance from the source, varying as the *inverse square of the distance*. Let us illustrate by an idealized case of a small source, regarded as a mathematical point, emitting S particles per second, the source "strength." As in Fig. 19.1, let the rate of flow through each unit of area of a sphere of radius R about the point be labeled ϕ(cm^{-2}-sec^{-1}). The flow through the whole sphere surface of area $4\pi R^2$ is then $\phi 4\pi R^2$, and if there is no intervening material, it can be equated to the source strength S. Then

$$\phi = \frac{S}{4\pi R^2}.$$

This relation expresses the inverse square spreading effect. If we have a surface covered with radioactive material or an object that emits radiation throughout its volume, the flux at a point of measurement can be found by addition of elementary contributions.

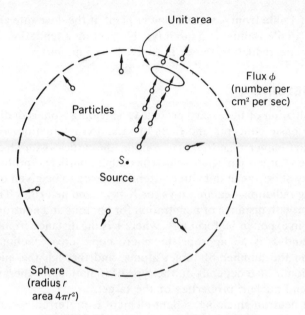

Fig. 19.1. Inverse square spreading of radiation.

Let us consider the neutron radiation at a large distance from an unreflected reactor operating at a power level of 1 MW. Since 1 W gives 3×10^{10} fissions/sec and the number of neutrons per fission is 2.5, the reactor produces 7.5×10^{16} neutrons/sec. Suppose that 20% of these escape and thus S is 1.5×10^{16} sec^{-1}. We apply the inverse square relation, neglecting attenuation in air, an assumption that would be correct for a reactor used to provide power for a spacecraft. Let us find the closest distance of safe approach to the reactor, i.e., where the neutron flux is below a safe level, say 0.15 cm^{-2}-sec^{-1}, as in Table 19.1. Solving the inverse-square formula for R, we obtain

$$ R = \sqrt{\frac{S}{4\pi\phi}} = \sqrt{\frac{1.5 \times 10^{16}}{(4\pi)(0.15)}} = 9 \times 10^{7} \text{ cm.} $$

This is a surprisingly large distance—about 560 miles. If the same reactor were on the earth, neutron attenuation in air would reduce this figure greatly, but the necessity for shielding by solid or liquid materials is clearly revealed by this calculation.

As another example, let us find how much radiation is received at a distance of 1 mile from a nuclear power plant, if the dose rate at the plant boundary, $\frac{1}{4}$-mile radius, is 5 mrem/yr. Neglecting attenuation in air, the inverse-square reduction factor is $\frac{1}{16}$ giving 0.03 mrem/yr.

19.4 SHIELDING

The evaluation of necessary protective shielding from radiation makes use of the basic concepts and facts of radiation interaction with matter described in Chapters 5 and 6. Let us consider the particles with which we must deal. Since charged particles—electrons, alpha particles, protons, etc.—have a very short range in matter, attention needs to be given only to the penetrating radiation—gamma rays (or X-rays) and neutrons. The attenuation factor with distance of penetration for photons and neutrons may be expressed in exponential form $e^{-\Sigma r}$, where r is the distance from source to observer and Σ is an appropriate macroscopic cross section. Now Σ depends on the number of target atoms, and through the microscopic cross section σ also depends on the type of radiation, its energy, and the chemical and nuclear properties of the target.

For fast neutron shielding, a light element is preferred because of the large neutron energy loss per collision. Thus hydrogenous materials such as water, concrete, or earth are effective shields. The objective is to slow neutrons within a small distance from their origin and to allow them to be absorbed at thermal energy. Thermal neutrons are readily captured by many materials, but boron is preferred because accompanying gamma rays are very weak.

Let us compute the effect of a water shield on the fast neutrons from the example reactor used earlier. The macroscopic cross section appearing in the exponential formula $e^{-\Sigma r}$ is now called a "removal cross section," since many fast neutrons are removed from the high-energy region by one collision with hydrogen, and eventually are absorbed as thermal neutrons. Its value for fission neutrons in water is around $0.10\ cm^{-1}$. A shield of thickness 8 ft = 244 cm would provide an attenuation factor of $e^{-24.4} = 10^{-10.6} = 2.5 \times 10^{-11}$. The inverse-square reduction with distance is

$$\frac{1}{4\pi r^2} = \frac{1}{4\pi (244)^2} = 1.3 \times 10^{-6}.$$

The combined reduction factor is 3.2×10^{-17}; and with a source of 1.5×10^{16} neutrons/sec, the flux is down to 0.5 neutrons/cm^2-sec, which is only slightly higher than the safe level. The addition of an extra foot of water shield would provide adequate protection, for steady reactor operation at least.

For gamma ray shielding, in which the main interaction takes place with atomic electrons, a substance of high atomic number is desired. Compton scattering varies as Z, pair production as Z^2, and the photoelectric effect as Z^5. Elements such as iron and lead are particularly useful for gamma shielding. The amount of attenuation depends on the material of the shield, its thickness, and the photon energy. The literature gives values of the "mass attenuation coefficient," which is the quotient of the macroscopic cross section and the material density. Typical values for a few elements at different energies are shown in Table 19.2. For 1 MeV gamma rays in iron, density 7.8 g/cm^3, we deduce that Σ is 0.467 cm^{-1}. For water, with $\frac{1}{9}$ g of hydrogen and $\frac{8}{9}$ g of oxygen per cubic centimeter, the mass attenuation coefficient is $(1/9)(0.126) + (8/9)(0.0637) = 0.0706$ cm^2/g; with density 1.0 g/cm^3, Σ is 0.0706 cm^{-1}. To achieve the same reduction in a beam of gammas, the thickness of an iron shield can be about $\frac{1}{6}$ that of a water shield.

As an example of gamma shielding calculations, we estimate the thickness of lead shield that should be provided for a point source of strength 1 millicurie (3.7×10^7 d/sec), emitting 1 MeV gamma rays, in order to bring the continuous exposure dose rate down to a level of 5 mrem/yr at the surface of the shield. From our previous calculations or Table 19.1, this corresponds to 0.33 cm^{-2}-sec^{-1}. We must take account of the fact that the simple exponential attenuation relation refers only to the transmission of gamma rays that have made no collision. Those which are scattered by the Compton effect can return to the stream and contribute to the dose, as sketched in Fig. 19.2. To account for this "buildup" of radiation, a *buildup factor B* depending on Σr is calculated. Figure 19.3 shows B for 1 MeV gammas in lead. The microscopic cross section for this radiation is found from Fig. 6.4 to be 24 barns. For lead with atomic

Table 19.2. Mass Attenuation Coefficients (cm²/g)

Energy (MeV)	Element			
	H	O	Fe	Pb
0.01	0.385	5.78	173	133
0.1	0.294	0.156	0.370	5.40
1	0.216	0.0637	0.0599	0.0708
2	0.0875	0.0446	0.0425	0.0455
10	0.0325	0.0209	0.0298	0.0484

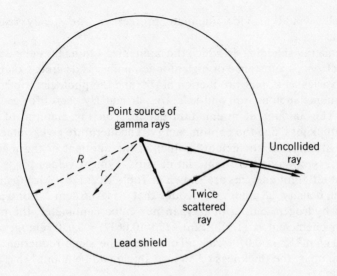

Point source of
gamma rays

Uncollided
ray

R

Twice
scattered
ray

Lead shield

Fig. 19.2. Buildup effect.

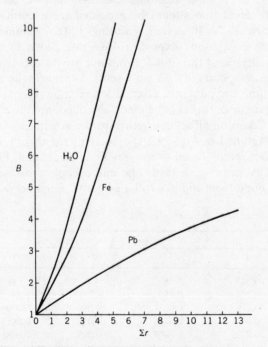

Fig. 19.3. Buildup factors, 1 MeV gammas.

weight $M = 207$ and density $\rho = 11.3 \, g/cm^3$, the atom number density N is $0.033 \times 10^{24} \, cm^{-3}$ and Σ is $0.80 \, cm^{-1}$. The combined effect of attenuation with material and distance, with buildup, may be written

$$\phi = \frac{BSe^{-\Sigma r}}{4\pi r^2}.$$

To find r, trial-and-error methods are required. The result is approximately 15 cm or 6 in., and the effective attenuation factor, as the ratio ϕ/S, is around 10^{-8}. The buildup factor turns out to be about 4.

Although calculations are performed in the design of equipment or experiments involving radiation, protection is ultimately assured by the measurement of radiation dosage. Portable detectors used as "survey meters" are available commercially. They employ the various detector principles described in Chapter 11, with the Geiger–Muller counter having the greatest general utility. Special detectors are installed to monitor general radiation levels or the amount of radioactivity in effluents.

The possibility of accidental exposure to radiation always exists in a laboratory or plant, in spite of all precautions. In order to have information immediately, personnel wear dosimeters, which are pen-size self-reading ionization chambers that detect and measure dose. For a more permanent record, film badges are worn. These consist of several photographic films of different sensitivity, with shields to select radiation types. They are developed periodically, and if significant exposure is noted, individuals are relieved of future work in areas with potential radiation hazards for a suitable length of time.

19.5 INTERNAL EXPOSURE

We now turn to the exposure of internal parts of an organism as a result of having taken in radioactive substances. Special attention will be given to the human body, but similar methods will apply to other animals and even to plants. Radioactive materials can enter the body by drinking, breathing, or eating, and to a certain extent can be absorbed through pores or wounds. The resulting dosage depends on many factors: (a) the amount that enters, which in turn depends on the rate of intake and elapsed time; (b) the chemical nature of the substance, which affects affinity with molecules of particular types of body tissue and which determines the rate of elimination (the term biological half-life is used in this connection, being the time for half of an initial amount to be removed); (c) the particle size, which relates to progress of the material through the body; (d) the radioactive half-life, the energy, and kind of radiation, which determine the activity and energy deposition rate, and the length of time the radiation exposure persists; (e) the radiosensitivity of the tissue, with the gastro-intestinal tract, reproductive organs, and bone marrow as the most important.

The maximum permissible concentration of an isotope in air or water can be calculated by consideration of all the above factors. Such an analysis in general is quite detailed, but there are some special cases in which a rough estimate can be made easily. An example is a gaseous fission product such as krypton, or the isotope tritium, half-life 12.3 yr, which is produced in a reactor in several ways. Suppose that a person is surrounded by air that contains a radioactive gas. If this "cloud" is large in comparison with radiation mean free paths or ranges, the rate of absorption of energy is equal to the rate of emission of energy. The beta particles of tritium have a low average energy, 0.006 MeV, and their range in air is very short.

Let us calculate the maximum permissible concentration (MPC) of tritium in air of density 1.29×10^{-3} g/cm^3, expressed as an activity per unit volume (μCi/cm^3), that will yield a dose rate of 5 mrem/yr. The energy released per second is

$$(\text{MPC } \mu\text{Ci/cm}^3)(3.7 \times 10^4 \text{ d/sec-}\mu\text{Ci})(0.006 \text{ MeV})(1.60 \times 10^{-13} \text{ J/MeV}),$$

while the energy absorbed per second is

$$\frac{(5 \times 10^{-3} \text{ rad/yr})(10^{-5} \text{ J/g-rad})(1.29 \times 10^{-3} \text{ g/cm}^3)}{3.16 \times 10^7 \text{ sec/yr}}.$$

Equating and solving, MPC $= 5.7 \times 10^{-8} \mu$Ci/cm^3. Although this corresponds to a very small number of particles in comparison with the 2.7×10^{19} cm^{-3} of air molecules, it turns out to be large in comparison with MPC values for more hazardous isotopes such as radium-226 or plutonium-239.

When there is more than one radioisotope present, the allowed concentrations must be limited. The criterion used is

$$\sum_i \frac{C_i}{(\text{MPC})_i} \leq 1$$

where i is an index of the isotope. This equation says the sum of quotients of actual concentrations and maximum permissible concentrations must be no bigger than 1. Values of MPC are tabulated in Code of Federal Regulations,

Title 10, Part 50, Licensing of Production and Utilization Facilities (frequently abbreviated "10CFR50").

19.6 RADIATION DAMAGE

Protection of living beings from the deleterious effects of radiation is of primary concern in the operation of any nuclear device, but attention must also be given to the possibility of radiation damage to nonliving materials. As discussed in Chapter 6, electrons, gamma rays, neutrons, and heavy charged particles can cause excitation, ionization, and dissociation of chemical substances, thus rendering them unsuitable for their original purpose. Generally speaking, organic materials such as plastics are sensitive to beta and gamma radiation, i.e., are most readily damaged. Their chemical bonds are relatively weak and easily broken. In contrast, metals are resistant to such radiation because their conduction electrons can absorb much of the energy of radiation without experiencing a structural change. However, metals are affected by neutron bombardment, principally through atom displacements. Thus metals in a nuclear reactor exhibit increases in hardness and tensile strength, with a loss in ductility. The effects depend on the dose, proportional to the total number of fast neutrons to which the material has been exposed. If the flux is $\phi = nv$, in a time t, the exposure is nvt, called the integrated flux or fluence. In the range of fluences 10^{19}–10^{21} per square centimeter most metals show such changes. The degree of damage tends to be smaller at higher material temperatures, since displaced atoms move about more readily and return to their original sites.

The success of the liquid metal fast breeder reactor will be determined in part by the ability of the fuel rods to withstand the rigorous thermal, mechanical, and radiation environment. Under the effects of radiation, the stainless steel cladding tends to swell and undergo creep, a slow stretching process, while the oxide fuel and its fission products interact with the cladding.

19.7 SUMMARY

Radiation protection of living organisms requires control of sources, barriers between sources and organism, or removal of the biological entity. Calculations required to evaluate external hazard include the dose as it depends on radiation flux and energy, material, and time; the inverse square spreading effect; and the exponential attenuation in shielding materials. Internal hazard depends on many physical and biological factors. Maximum permissible concentrations of radioisotopes in air or water can be deduced from the properties of the emitter and the specified dose rate limits. Radiation damage in nonliving substances is of concern, since organic chemicals are readily affected, and neutron bombardment changes reactor metal properties.

19.8 PROBLEMS

19.1. What is the rate of exposure in mrem/yr corresponding to a continuous gamma ray flux of 100 cm^{-2}-sec^{-1}? What dose equivalent would be received by a person who worked 40 hr/wk throughout the year in such a flux?

19.2. A Co-60 source is to be selected to test radiation detectors for operability. Assuming that the source can be kept at least 1 m from the body, what is the largest strength acceptable (in μCi) to assure an exposure rate of less than 500 mrem/yr? (Note that two gammas of energy 1.1 and 1.3 MeV are emitted.)

19.3. By comparison with the tritium analysis, estimate the MPC in air for Kr-85, average beta particle energy 0.22 MeV.

19.4. The nuclear reactions resulting from thermal neutron absorption in boron and cadmium are

$$^{10}_{5}B + ^{1}_{0}n \rightarrow ^{7}_{3}Li + ^{4}_{2}He,$$

$$^{113}_{48}Cd + ^{1}_{0}n \rightarrow ^{114}_{48}Cd + \gamma(5\ MeV).$$

Which material would you select for a radiation shield? Explain.

19.5. Find the gamma ray flux that gets through a spherical lead shield of 12 cm radius if the source of 1 MeV gammas is of strength 200 mCi.

19.6. The maximum permissible concentration (MPC) of unidentified beta- and gamma-emitting isotopes in water is 10^{-7} μCi/cm^3. In order to assure that the actual release is no more than 1% of the MPC, a limitation on the discharge rate (r) in gallons per minute (gpm) must be applied for each radioactive solution of specific activity c (μCi/cm^3). Assuming a further dilution of any fluid released by a river stream flow of 1500 gpm, develop a working formula relating r to c, and plot a graph for convenient use. Suggestion: 3-cycle log–log paper.

19.7. Water discharged from a nuclear plant contains in solution traces of strontium-90, cerium-144, and cesium-137. Assuming that the concentrations of each isotope are proportional to their fission yields, find the allowed activities per ml of each. Note the following data:

Isotope	Half-life	Yield	MPC(μCi/ml)†
^{90}Sr	28.8 y	0.0575	1×10^{-5}
^{144}Ce	284.5 d	0.0545	3×10^{-4}
^{137}Cs	30.2 y	0.0611	3×10^{-3}

†According to 10CFR50.

20

Reactor Safety

It is well known that the accumulated fission products in a reactor that has been operating for some time constitute a potential source of radiation hazard. Assurance is needed that the integrity of the fuel is maintained throughout the operating cycle, with negligible release of radioactive materials. This implies limitations on power level and temperature, and adequacy of cooling under all conditions. Fortunately, inherent safety is provided by physical features of the fission chain reaction. In addition, the choice of materials, their arrangement, and restrictions on modes of operation give a second level of protection. Devices and structures that minimize the chance of accident and the extent of radiation release in the event of accident are a third line of defense. Finally, nuclear plant location at a distance from centers of high population density results in further protection.

We shall now describe the dependence of numbers of neutrons and reactor power on the multiplication factor, which is in turn affected by temperature and control rod absorbers. Then we shall examine the precautions taken to prevent release of radioactive materials to the surroundings and discuss the philosophy of safety.

Thanks are due Robert M. Koehler of Duke Power Company for suggestions on parts of this chapter.

20.1 NEUTRON POPULATION GROWTH

The multiplication of neutrons in a reactor can be described by the effective multiplication factor k, as discussed in Chapter 12. The introduction of 1 neutron produces k neutrons, they in turn produce k^2, and so on. Such a behavior would seem to be completely analogous to the increase in principal with compound interest or the exponential growth of human population. We shall see shortly that there are some important differences, but it will be desirable to develop an expression for the growth with time for *any* population n, if for each individual the gain per cycle of reproduction is $\delta k = k - 1$ and the time for 1 cycle is l. In an infinitesimal time dt, the gain with n individuals is dn, and by proportions

$$\frac{dn}{dt} = \frac{\delta k n}{l}.$$

If the coefficient of n on the right side is constant, and if the number present at time zero is n_0, the solution is found to be

$$n = n_0 e^{t/T},$$

where T is the "period," the time for the population to increase by a factor $e = 2.718\ldots$, given by $T = l/\delta k$. When applied to people, the formula states that the population grows more rapidly the more frequently reproduction occurs and the more abundant the progeny.

A typical cycle time l for neutrons in a thermal reactor is very short, around 10^{-5} sec, so that a δk as small as 0.02 would give a very short period of 0.0005 sec. The growth according to the formula would be exceedingly rapid, and if sustained would consume all of the atoms of fuel in a fraction of a second.

A peculiar and fortunate fact of nature provides an inherent reactor control for values of δk in the range 0 to around 0.0065. Recall that around 2.5 neutrons are released from fission. Of these, some 0.65% appear later as the result of radioactive decay of a certain group of fission products, and are thus called *delayed neutrons*. The average half-life of the isotopes from which they come, taking account of their yields, is around 8.8 sec. This corresponds to a mean life $\tau = t_H/0.693 = 12.7$ sec, as the average length of time required for a radioactive isotope to decay. Although there are very few delayed neutrons, their presence extends the cycle time greatly and slows the rate of growth of neutron population. To understand this effect, let β be the fraction of all neutrons that are delayed, a value 0.0065 for U-235; $1 - \beta$ is the fraction of those emitted instantly as "prompt neutrons." If the length of time before the delayed neutrons appear is τ, but the prompt neutrons appear instantly, the average delay is $\beta \tau + (1 - \beta) 0 = \beta \tau$. Now since $\beta = 0.0065$ and $\tau = 12.7$ sec, the product is 0.083 sec, greatly exceeding the multiplication cycle time, which is only 10^{-5} sec. The delay time can thus be regarded as the effective generation time, $\bar{l} = \beta \tau$. This approximation holds for values of δk much less than β.

For example, let $\delta k = 0.001$, and use $\bar{l} = 0.083$ sec in the exponential formula. In one second $n/n_0 = e^{0.012} = 1.01$, a very slight increase.

On the other hand, if δk is greater than β we still find very rapid responses, even with delayed neutrons. If all neutrons were prompt, 1 neutron would give a gain of δk, but since the delayed neutrons actually appear much later, they cannot contribute to the immediate response. The apparent δk is then $\delta k - \beta$, and the cycle time is l. We can summarize by listing the period T for the two regions.

$$\delta k \ll \beta, \qquad T \simeq \frac{\beta \tau}{\delta k},$$

$$\delta k \gg \beta, \qquad T \simeq \frac{l}{\delta k - \beta}.$$

Even though β is a small number, it is conventional to consider δk small only if it is less than 0.0065 but large if it is greater. Figure 20.1 shows the

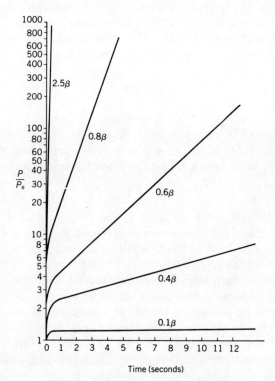

Fig. 20.1. Effect of delayed neutrons.

growth in reactor power for several different values of reactivity ρ, defined as $\delta k/k$. Since k is close to 1, $\rho \simeq \delta k$. We conclude that the rate of growth of neutron population or reactor power is very much smaller than expected, so long as δk is kept well below the value β, but that rapid growth will take place if δk is larger than β.

We have used the value of β for U-235 for illustration, but should note that its effective value depends on reactor size and type of fuel, e.g., β for Pu-239 is only 0.0021. Also, the value of neutron cycle time depends on the energy of the predominant neutrons. The l for a fast reactor is much shorter than that for a thermal reactor.

20.2 TEMPERATURE EFFECTS

Reactor safety is enhanced by certain inherent temperature effects in the fuel. An increase in power produced by an applied reactivity tends to cause a temperature increase that gives rise to a negative reactivity, which cancels the initial value. In certain reactors the effect is due to expansion that results in larger neutron mean free paths and hence greater leakage. This response may be described in terms of a negative temperature coefficient of reactivity α, with reactivity proportional to temperature change

$$\rho = \alpha \Delta T.$$

For example, if α is $-10^{-4}/°C$, a temperature rise of 50°C could give a reactivity of -0.005, which is almost as large as β.

One can also define a negative power coefficient labeled a, i.e., the reactivity may be written where $\Delta P/P$ is the fractional increase in power.

$$\rho = a\Delta P/P$$

For a typical PWR, with its large U-238 content, the coefficient a is $- 0.012$ such that a 1% change in power would give rise to a reactivity of 0.00012. Figure 20.2 shows the trend of power with time in a reactor that has this self-regulation. Instead of continuing to rise as in Figure 20.1 in which no temperature effects were considered, the power flattens out and becomes constant. This negative feedback goes by the name of Doppler effect, by analogy to frequency changes in sound or light when there is a relative motion of the source and observer. The thermal agitation of fuel atoms increases with the temperature and thus the effective resonance cross section for neutrons colliding with the nuclei is increased.

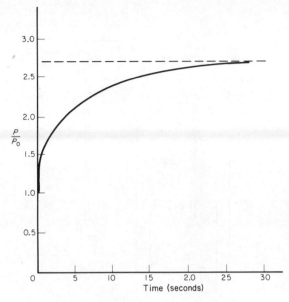

Fig. 20.2.   Effect of temperature on power.

20.3 SAFETY THROUGH DESIGN, FABRICATION, AND OPERATION

Even though a reactor is relatively insensitive to increases in multiplication in the region below β and temperature increases tend to shut the reactor down, additional protection is provided in the design and through operation practices.

Part of the control of a reactor of the PWR type is provided by boron solution (see Section 13.4). This "chemical shim" balances the excess fuel loading and is adjusted as fuel is consumed during the operating life. In addition, PWR reactors are equipped with several groups of movable rods of neutron absorbing material, as shown in Fig. 20.3. The rods serve three main purposes: (a) to permit temporary increases of multiplication that bring the reactor up to the desired power level or to make adjustments to the power; (b) to cause changes in the flux and power shape within the core, usually striving for uniformity; and (c) to shut down the reactor automatically or manually in the event of unusual behavior. To ensure the effectiveness of the shutdown role, several groups of "safety rods" are kept withdrawn from

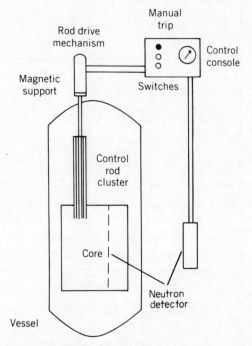

Fig. 20.3. Reactor control.

the reactor at all times during operation. In the PWR, they are supported by electromagnets that release the rod on interruption of current while in the BWR they are driven in from the bottom of the vessel by hydraulic means.

An instrumentation system is provided to detect an excessive neutron flux and thus power level, to provide signals calling for a "trip" of the reactor. As sketched in Fig. 20.3, independent detectors are located both inside the core and outside the reactor vessel. Data from core detectors are processed by a computer to determine whether or not power distributions are acceptable.

Since almost all of the radioactivity generated by a reactor appears in the fuel elements, great precautions are taken to assure the integrity of the fuel. Care is taken in fuel fabrication plants to produce fuel pellets that are identical chemically, of the same size and shape, and of common U-235 concentration. If one or more pellets of unusually high fissile material content were used in a reactor, excessive local power production and temperature would result. The metal tubes that contain the fuel pellets are made sufficiently thick to stop the fission fragments, to provide the necessary mechanical strength to support the column of pellets, and to withstand erosion by water flow or corrosion of water at high temperatures. Also, the tube must sustain a variable pressure difference caused by moderator-coolant outside and fission product gases inside. The "cladding" material usually selected for low neutron absorption and for resistance to chemical action, melting, and radiation damage in thermal reactors is zircaloy, an alloy that is about 98% zirconium with small amounts of tin, iron, nickel, and chromium. The tube is formed by an extrusion process that eliminates seams, and special fabrication and inspection techniques are employed to assure that there are no defects such as deposits, scratches, holes, or cracks.

Each reactor has a set of specified limits on operating parameters to assure protection against events that could cause hazard. Typical of these is the upper limit on total reactor power, which determines temperatures throughout the core. Another is the ratio of peak power to average power which is related to hot spots and fuel integrity. Protection is provided by limiting the allowed control rod position, reactor imbalance (the difference between power in the bottom half of the core and the top half) and reactor tilt (departure from symmetry of power across the core), maximum reactor coolant temperature, minimum coolant flow, and maximum and minimum primary system pressure. Any deviation causes the safety rods to be inserted to trip the reactor. Maintenance of chemical purity of the coolant to minimize corrosion, limitation on allowed leakage rate from the primary cooling system, and continual observations on the level of radioactivity in the coolant serve as further precautions against release of radioactive materials.

20.4 STANDARDS, QUALITY ASSURANCE, LICENSING, AND REGULATION

In the foregoing paragraphs we have alluded to a few of the physical features and procedures employed in the interests of safety. These have evolved from experience over some twenty years, as described in Chapter 17, and much of the design and operating experience has been translated into widely used *standards*, which are descriptions of acceptable practice. Professional technical societies, industrial organizations, and the federal government cooperate in the development of these useful documents.

In addition, requirements related to safety have a legal status, since all safety aspects of nuclear systems are rigorously regulated by federal law, administered by the United States Nuclear Regulatory Commission (NRC). Before a prospective owner of a nuclear plant can receive a permit to start construction, he must submit a comprehensive preliminary safety analysis report (PSAR) and an environmental impact statement. Upon approval of these, a final safety analysis report (FSAR), technical specifications, and operating procedures must be developed in parallel with the manufacture and construction. An exhaustive testing program of components and systems is carried out at the plant. The documents and test results form the basis for an operating license.

Throughout the analysis, design, fabrication, construction, testing and operation of a nuclear facility, adequate *quality control* (QC) is required. This consists of a careful documented inspection of all steps in the sequence. In addition, a *quality assurance* (QA) program that verifies that quality control is being exercised properly is imposed. Licensing by the NRC is possible only if the QA program has satisfactorily performed its function. During the life of the plant, periodic inspections of the operation are made by the NRC to ascertain whether or not the owner is in compliance with safety regulations, including commitments made in Technical Specifications and the FSAR.

20.5 EMERGENCY CORE COOLING

The design features and operating procedures for a reactor are such that under normal conditions a negligible amount of radioactivity will get into the coolant and find its way out of the primary loop. Knowing that abnormal conditions can exist, the worst possible event, called a design basis accident, is postulated. Backup protection equipment, called engineered safety features, is provided to render its effect negligible. A loss of coolant accident (LOCA) is the condition typically assumed, in which the main coolant piping somehow breaks and thus the pumps cannot circulate coolant through the core. Although the reactor power would be reduced immediately by use of safety rods in such a situation, there is a continuing supply of heat from the decaying fission products that would tend to increase temperatures above the melting point of the fuel and cladding. In a severe situation, the fuel tubes would be damaged, and a considerable amount of fission products released. In order to prevent melting, an emergency core cooling system (ECCS) is provided in water-moderated reactors, consisting of auxiliary pumps that inject and circulate cooling water to keep temperatures down. The operation of a typical ECCS can be understood by study of some schematic diagrams.

The basic reactor system (Fig. 20.4) includes the reactor vessel, the primary coolant pump, and the steam generator, all located within the containment building. The system actually may have more than one steam generator and pump—these are not shown for ease in visualization. We show in Fig. 20.5 the auxiliary equipment that constitutes the engineered safety (ES) system. First is the *high-pressure injection system*, which goes into operation if the vessel pressure expressed in pounds per square inch (psi) drops from a normal value of around 2250 psi to about 1500 psi as the result of a small leak. Water is taken from the borated water storage tank

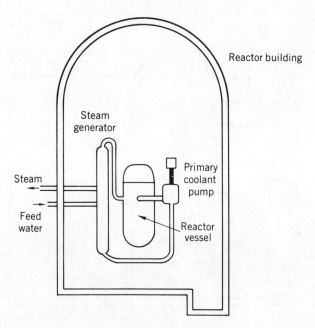

Fig. 20.4.   Reactor containment.

and introduced to the reactor through the inlet cooling line. Next is the *core flooding tank*, which delivers borated water to the reactor through separate nozzles in the event a large pipe break occurs. Such a rupture would cause a reduction in vessel pressure and an increase in building pressure. When the vessel pressure becomes around 600 psi the water enters the core through nitrogen pressure in the tank. Then if the primary loop pressure falls to around 500 psi, the *low-pressure injection* pumps start to transfer water from the borated water storage tank to the reactor. When this tank is nearly empty, the pumps take spilled water from the building sump as a reservoir and continue the flow, through coolers that remove the decay heat from fission products. Another feature, the building spray system, also goes into operation if the building pressure increases above about 4 psi. It takes water from the borated water storage tank or the sump and discharges it from a set of nozzles located above the reactor, in order to provide a means for condensing steam. At the same time, the reactor building emergency cooling units are operated to reduce the temperature and pressure of any released vapor, and reactor building isolation valves are closed on unnecessary piping to prevent the spread of radioactive materials outside the building.

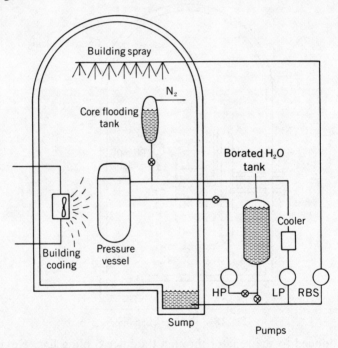

Fig. 20.5. Emergency core cooling system.

We can estimate the magnitude of the problem of removing fission product heat. For a reactor fueled with U-235, operated for a long time at power P_0 and then shut down, the power associated with the decay of accumulated fission products is $P_f(t)$, given by an empirical formula such as

$$P_f(t) = P_0 A t^{-a}.$$

For times larger than 10 sec after reactor shutdown the decay is represented approximately by using $A = 0.066$ and $a = 0.2$. We find that at 10 sec the fission power is 4.2% of the reactor power. By the end of a day, it has dropped to 0.68%, which still corresponds to a sizeable power, viz., 20 MW for a 3000 MWt reactor. The ECCS must be capable of limiting the surface temperature of the zircaloy cladding to specified values, e.g., 2200°F, of preventing significant chemical reaction, and of maintaining cooling over the long term after the postulated accident.

The role of the steel-reinforced concrete reactor building is to provide containment of fission products that might be released from the reactor. It is designed to withstand internal pressures and to have a very small leak rate. The reactor building is located within a zone called an exclusion area, of radius of the order of half a kilometer, and the nuclear plant site is several kilometers from any population center.

A series of experiments called Loss of Flow Tests (LOFT) has been done at Idaho Falls to check the adequacy of mathematical models and computer codes related to LOCA/ECCS. A double-ended coolant pipe break can be introduced and the ability to inject water against flow reversal and water vapor can be determined. Preliminary tests in 1979 showed that peak temperatures reached were lower than predicted, indicating conservatism in the calculation methods.

The results of an extensive investigation of reactor safety were published in 1976, involving sixty scientists and engineers and several million dollars expense. Assessed were the chances of failure of nuclear reactors and their auxiliary equipment, the possibility of failure of emergency equipment to function, and finally the consequences of a core meltdown with the release of radioactivity. The conclusion reached in this Safety Report, or "Rasmussen Report" after its principal author, was that the chance of a member of the public being harmed was extremely low in comparison with the chance for natural or accepted man-made risks. The report has received some criticism for vagueness and optimism. The main difficulty is that accident experience is too small to form a basis for statistical estimates.

20.6   THREE MILE ISLAND

On March 28, 1979 an accident occurred at a reactor called Three Mile Island (TMI) near Harrisburg, Pennsylvania. A small amount of radio-activity was released, and a number of people were evacuated or left the area for a while. The event was reported fully by news media and caused alarm throughout the region and beyond. In view of the great public interest in the incident and the consequent potential effect on the growth of nuclear power, we shall attempt to describe what happened at TMI. Then we shall suggest some implications of the events.

In Sections 13.9 and 14.4 we have described the features of a typical pressurized water reactor system. We shall refer especially to Figs. 14.6 and 14.7 in reviewing the TMI chronology. The reactor was operating steadily at nearly full power when at 4 a.m. there was a malfunction in the steam generator's feedwater system. (Recall that the feedwater pump returns the condensed steam from the turbine.) Because of this failure, the turbine generator was automatically tripped and control rods were driven into the reactor to reduce its power. To this point, nothing unusual had happened. Three backup feedwater pumps should have provided the necessary water. However, they could not because, as it was later learned, a valve to the steam generator had been left closed by mistake. Not until some 8 minutes was this discovered and the valve opened. As a result, the steam generators dried out. Thus the primary water coolant temperature and pressure increased to about 2355 psi, causing a relief valve on the pressurizer to open. The coolant then could escape to a vessel called the quench tank designed to condense and cool any releases from the reactor system. The pressurizer relief valve stuck open, a fact not realized by the operators for 2 hours. Therefore a considerable amount of coolant was released, eventually filling the quench tank and causing a rupture disk on the tank to blow out. Coolant water containing some radioactivity spilled into the containment building, finding its way to the sump. In the meantime, the reactor pressure continued to fall. At 1600 psi, the emergency core cooling system actuated, as it was supposed to. The high-pressure pumps injected makeup water into the reactor vessel. According to the observations made by the operators, the pressurizer appeared to be filled with water, a condition that would prevent its functioning. They decided to shut off the emergency cooling system and later to stop the main reactor coolant pumps. This severe lack of water caused the core to heat up and become uncovered. Although the main fission power had been cut off, there remained the large amount of residual heat from the decaying fission products. The coolant flow in the core was inadequate to cool the fuel rods and some damage was experienced. Considerable radioactivity, especially of noble gases such as xenon and krypton, along with iodine, was transferred out of the reactor. The design of the system was such that sump pumps automatically sent the radioactive water from the containment into tanks in an auxiliary building next door. The tanks overflowed, permitting radioactive material to escape through filters into the atmosphere. In the course of trying to get water back into the containment building, additional releases were made. Back at the reactor, the cooling system was finally turned on and the core temperature began to fall. However, there was evidence that metal–water reactions had caused hydrogen to be evolved, and it was believed that a large bubble of potentially explosive gas had been formed in the top of the reactor vessel. Efforts were directed for several days toward eliminating this. It is not certain that such a bubble actually existed. Soon after the release of radioactive gases, measurements of atmospheric contamination were initiated by detectors in an airplane, a truck, and at fixed locations in the vicinity. The best estimates are that the highest possible dose to anyone was less than 100 mrem. This was based on assumed continuous exposure outdoors at the site boundary for 11 days. The average exposure to people within 50 miles was estimated to be only 11 mrem, noted to be less than that due to a medical X-ray. As a result of a warning by the governor of Pennsylvannia, many people, especially pregnant women, left the area for several days. Estimates published by the Department of Health, Education and Welfare indicate that the exposure over the lifetimes of the two million people in the region there would be statistically only one additional cancer death (out of 325,000 due to other causes).

We can suggest some implications of the Three Mile Island incident. It is impossible to find the exact causes of all the various problems. A number of reasonable conclusions can be drawn, however, and the possible consequences of the incident can be assessed. The TMI accident was the result of a combination of design deficiency, equipment failure, and operator error. In the design area, it should not have been possible for radioactive water to be pumped out of the containment without anyone's knowledge. Also, instrumentation to allow operators full knowledge of the system thermal-hydraulic status should have been available. The main equipment failure was the stuck pressurizer valve. In this incident the equipment as a whole performed quite well, but there are many examples of failure of valves, pumps, and switches that could be eliminated by better quality control during fabrication and by better inspection and maintenance. Operator errors were numerous, including the closing of the valve in the feedwater line, misreading the condition of the pressurizer, and shutting off both the emergency core cooling pumps and the reactor cooling pumps.

The consequences of the event for the future depend on one's point of view. Opponents of nuclear power view it as proof of their contention that protection of the public cannot be assured and thus all reactors should be shut down or new construction stopped, at least. Supporters of nuclear power point out that no one was injured in the TMI affair, that the emergency equipment functioned, that the reactor core stood up better than expected under abuse, and that the experience will prompt new safety precautions and improved operator training.

20.7 PHILOSOPHY OF SAFETY

The subject of safety is a subtle combination of technical and psychological factors. Regardless of the precautions that are provided in the design, construction, and operation of any device or process, the question can be raised "Is it safe?". The answer cannot be a categorical "yes" or "no," but must be expressed in more ambiguous terms related to the chance of malfunction or accident, the nature of protective systems, and the consequences of failure. This leads to more philosophical questions such as "How safe is safe?" and "How safe do we want to be?".

Every human endeavor is accompanied by a certain risk of loss or damage or hazard to individuals. In the act of driving an automobile on the highways, or in turning on an electrical appliance in the home, or even in the process of taking a bath, one is subject to a certain danger. Everybody agrees that the consumer deserves protection against hazard outside his personal control, but it is not at all clear as to what lengths it is necessary to go. In the absurd limit, for instance, a complete ban on all mechanical conveyances would assure that no one would be killed in accidents involving cars, trains, airplanes, boats, or spacecraft. Few would accept the restrictions thus implied. It is easy to say that reasonable protection should be provided, but the word "reasonable" has different meanings among people. The concept that the benefit must outweigh the risk is appealing, except that it is very difficult to assess the risk of an innovation for which no experience or statistical data are available, or for which the number of accidents is so low that many years would be required for adequate statistics to be accumulated. Nor can the benefit be clearly defined. A classic example is the use of a pesticide that assures protection of the food supply for many, with finite danger to certain sensitive individuals. To the person affected adversely, the risk completely overshadows the benefit. The addition of safety measures is inevitably accompanied by increased cost of the device or product, and the ability or willingness to pay for the increased protection varies widely among people.

It is thus clear that the subject of safety falls within the scope of the social–economic–political structure and processes and is intimately related to the fundamental conflict of individual freedoms and public protection by control measures. It is presumptuous to demand that every action possible should be taken to provide safety, just as it is negligent to contend that because of evident utility, no effort to improve safety is required. Between these extreme views, there remains an opportunity to arrive at satisfactory solutions, applying technical skill accompanied by responsibility to assess consequences. It is most important to provide understandable information, on which the public and its representatives can base judgments and make wise decisions as to the proper level of investment of effort and funds.

20.8 SUMMARY

Prevention of release of radioactive fission products and fuel isotopes is the ultimate purpose of safety features. Inherent reactor safety is provided by delayed neutron and temperature effects. Control rods permit quick shutdown, and reactor components are designed and constructed to minimize the chance of failure, with licensing administered by a federal agency. Equipment is installed to reduce the hazard in the event of a postulated accident. The Three Mile Island incident served to stimulate increased attention to safety. Controversy about nuclear energy is related to safety and the relation of benefit and risk.

20.9 PROBLEMS

20.1. (a) If the total number of neutrons from fission by thermal neutrons absorbed in U-235 is 2.43, how many are delayed and how many are prompt?

(b) A reactor is said to be "prompt critical" if it has a positive reactivity of β or more. Explain the meaning of the phrase.

(c) What is the period for a reactor with neutron cycle time 5×10^{-6} sec if the reactivity is 0.013?

(d) What is the period if instead the reactivity is 0.0013?

20.2. A reactor is operating at a power level of 250 MWe. Control rods are removed to give a reactivity of 0.0005. Noting that this is much less than β, calculate the time required to go to a power of 300 MWe, neglecting any temperature feedback.

20.3. When a large positive reactivity is added to a fast reactor assembly, the power rises to a peak value and then drops, crossing the initial power level. In this response, which is the result of a negative temperature effect, the times required for the rise and fall are about the same. If the neutron cycle time is 4×10^{-6} sec, what would be the approximate duration of an energy pulse resulting from a reactivity of 0.0165, if the peak power is 10^3 times the initial power? See figure.

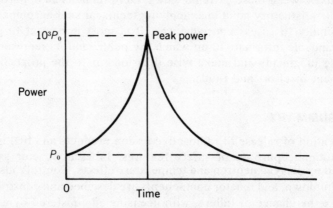

20.4. During a "critical experiment," in which fuel is initially loaded into a reactor, a fuel element of reactivity worth 0.0036 is suddenly dropped into a core that is already critical. If the temperature coefficient is $-9 \times 10^{-5}/°C$, how high will the temperature of the system go above room temperature before the positive reactivity is canceled out?

20.5. How long will it take for a fully withdrawn control rod in a reactor of height 4 m to drop into a reactor core neglecting all friction and buoyancy effects? (Recall $s = \frac{1}{2}gt^2$ with $g = 9.8$ m/sec².)

20.6. Calculate the ratio of fission product power to reactor power for four times after shutdown—1 day, 1 week, 1 month, and 1 year, using the approximation $A = 0.066$, $a = 0.2$.

20.7. A reactivity of –0.0025 due to Doppler effect results when the thermal power goes from 2500 MW to 2800 MW. Estimate the contribution of this effect on the power coefficient for the reactor.

21

Radioactive Waste Processing and Disposal

The byproducts of every neutron-induced fission of a heavy element are two lighter fragments that are usually highly radioactive, with half-lives ranging from fractions of a second to thousands of years. There being no practical way in sight to render these fission products nonradioactive and therefore inert, we face the fact that the use of nuclear energy is accompanied by a continuing demand for safe handling, transportation, processing, storage, and disposal of potentially hazardous materials. Each of these steps will be reviewed in this chapter.

21.1 AMOUNTS OF FISSION PRODUCTS

An appreciation can be gained of the magnitude of the problem of handling reactor-produced radioactive materials by a study of their physical characteristics. First, we note that the weight and volume of fission products are rather small. When a U-236 nucleus splits, the mass-energy released is only $200 \text{ MeV}/(931 \text{ MeV/amu}) \simeq 0.2$ amu. The atomic masses of fission products still add up to approximately 236, and thus we can essentially equate the weights of fuel fissioned and waste products. For each megawatt-day of energy release, 1.3 g of U-235 are consumed. Of this, 86% is fission, so the amount of U-235 fissioned is $(0.86)(1.3) = 1.1 \text{ g/MWd}$.

A reactor operating at 3000 MWt thus produces 3.3 kg of fission products per day or about 1200 kg per year. If we assume an average specific gravity of 10, i.e., 10^4 kg/m^3, the volume comes out to be only $0.12 \text{ m}^3/\text{yr}$. This corresponds to a cube 0.5 m on a side. The amount of material that must be handled varies with the stage of processing, starting with the removal of the spent-fuel assemblies and ending with the complete separation of useful fuels and wastes. On the other hand, the decay of short-lived isotopes reduces the amount of radioactive material.

Although the actual amount of material is small, the heat generation rate, the activity, and the resultant radiation level is high for a considerable time after fuel is removed from the reactor. We may again apply the decay heat formula (Chapter 20) to estimate the requirements on cooling and shielding. The power from decay at a time as long as 3 months $(7.8 \times 10^6 \text{ sec})$ after shutdown of a 3000-MW reactor is $P = (3000)(0.066)(7.8 \times 10^6)^{-0.2} = 8.3$ MW. If we assumed that typical particles released have an energy of 1 MeV, this would correspond to 1.4×10^9 Ci.

21.2 FUEL HANDLING

Over the years since World War II, the production of fissile material for U.S. weapons has involved chemical processing to separate the desired plutonium. The residual fission products have been stored in aqueous form in underground tanks, mainly at the Hanford site in the state of Washington and at the Savannah River Laboratory in South Carolina.

In contrast, relatively little of the spent fuel from nuclear power plants has been reprocessed — first for lack of adequate facilities and more recently as a matter of policy. In this section we shall review the handling of fuel at the reactor site and long-term storage of irradiated assemblies. Later, we discuss spent fuel reprocessing and various modes of recycling materials.

The mechanical aspects of reactor fuel handling are straightforward. At the end of an operating period, of the order of a year in a typical light water reactor, the head of the reactor vessel is removed and set aside. The whole space above the vessel is then filled with water to allow fuel assemblies to be removed while immersed. By means of movable hoists the individual fuel assemblies weighing about 600 kg (1320 lb) are extracted from the reactor and transferred to a water-filled storage pool in an adjacent building. Part of the fuel assemblies remaining in the reactor are moved from one position in the reactor to another, and fresh fuel assemblies are brought from storage to fill vacant spaces. Typically one-third of the fuel (sixty assemblies for a PWR) is removed from the reactor, and one-third new fuel is added. The internal fuel arrangement at the start of a cycle is selected to minimize local heating and to optimize burnup and energy production. The principal hazard in handling fuel assemblies is due to gamma radiation from the great variety of isotopic species that comprise the fission products. Water immersion is needed both for protective shielding and removal of decay heat energy.

The storage facilities at a reactor plant include vertical racks composed of stainless steel that support and separate fuel assemblies in a water bath. The multiplication factor k of an individual fuel element is fairly close to 1. Thus in order to prevent accidental criticality in the storage pool, it is necessary to maintain several centimeters distance between assemblies and to include neutron absorbing material such as boron. Control of the purity of water and the temperature of fuel is provided by means of filters, demineralizers, and coolers. Over the period of storage at the reactor facility, the radioactivity declines to values safe for transportation of the assemblies to some off-site storage location or to a reprocessing plant.

The fuel that comes out of the reactor has been irradiated for three years in the reactor, during which time much of the U-235 has been consumed. A small amount of U-238 has been burned to create Pu-239, part of which has been used to help produce power and part to yield higher mass is isotopes Pu-240, the fissile Pu-241, and Pu-242. Figure 21.1 shows the change in isotopic composition of the fuel as the result of irradiation for three years in a flux of 3×10^{13}/cm²-sec. The quantities in parentheses are percentages (in terms of original uranium content). We see that U with U-235 in amount 3.3% is converted into a mixture of fission products 3.5%, U-235 0.8%, and several isotopes of plutonium.

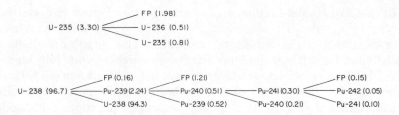

Fig. 21.1.   Isotopic composition of typical LWR fuel before and after exposure (percentages in parentheses). FP = fission products.

This particular calculation ended with the production of Pu-242. In fact some higher transuranic isotopes such as americium and curium are also produced in small but significant quantities. The weights of each fuel isotope can be calculated from the fact that the U per assembly is around 469 kg. Thus the amount of Pu-239 is (0.0052)(469 kg) = 2.44 kg. This isotope is hazardous because of its alpha particle emission with half-life 24,131 years. Let us calculate the amounts of fissile fuel before and after irradiation. In fresh fuel we have per assembly a weight of U-235 of (0.033)(469) = 15.48 kg. Afterwards the fissile weights are 3.80 kg of U-235, 2.44 kg of Pu-239, and 0.47 kg of Pu-241, a total of 6.67 kg, i.e., 43% as much as at the start. This constitutes a potentially valuable energy source, but one that can be realized only by reprocessing of the spent fuel.

By national policy established in 1977, reprocessing of spent fuel to separate fuels from fission products was stopped in the United States. The basis for the decision by President Carter was that reprocessing increased the likelihood that plutonium would be extracted illegally for use in nuclear weapons. The "once-through" fuel cycle was adopted in which spent fuel is set aside and not reprocessed. Plans were thus made by the utilities using reactors to store the fuel in water-filled pools at the plant sites until facilities away from the reactor (AFR) were available.

In the subsequent sections we shall discuss transportation of radioactive materials as is required to distribute spent fuel among available storage facilities or to carry selected fuel assemblies to reprocessing facilities for test purposes.

21.3 TRANSPORTATION

Because of residual radioactivity, spent fuel is shipped in casks that protect against radiation exposure of workers and the public, the release of radioactivity, and accidental criticality. The shipping containers sketched in Fig. 21.2(a) consist of steel tanks weighing when fully loaded with 7 PWR assemblies up to 64,000 kg (70 tons), of length 5 m (16.5 ft) and diameter 1.5 m (5 ft). The casks contain boron tubes to prevent criticality, heavy metal for gamma-ray shielding, and water for cooling and additional shielding. The vessel is sealed to prevent the escape of radioactive materials; fins on the outside of the cask help remove heat during shipment. A portable air-cooling system is attached when the cask is loaded

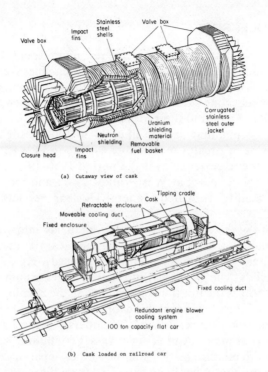

(a) Cutaway view of cask

(b) Cask loaded on railroad car

Fig. 21.2.   Spent fuel shipping cask. (Courtesy of General Electric Company.)

on a railroad car as in Fig. 21.2(b). The cask is designed to withstand normal conditions involving temperature variations, wetting, vibrations, and shocks.

By regulation, the cask is also required to withstand in sequence a 30-ft (~ 10 m) free fall onto a solid surface, a 40-in. (~ 1 m) fall to strike a 6-in. (~ 15 cm) diameter pin, a 30-minute exposure to a fire at temperature 1475°F (~ 800°C), and complete immersion for a period of 8 hours in 3 ft of water. The specifications are intended to simulate real conditions in road accidents. To further test the integrity of similar casks, a tractor-trailer carrying one was made to collide at 37 km/hr (60 mi/hr) into a concrete wall. The vehicle was demolished but the cask was merely scratched.

In the next section we describe the mechanical and chemical treatment of spent fuel as was envisaged by the U.S. nuclear industry prior to governmental action in 1977 or that practiced in various foreign countries such as Great Britain, France, Germany, and Japan.

21.4 REPROCESSING

From the standpoint of economics and resources, spent fuel contains a sufficient amount of fissile material to warrant chemical reprocessing to eliminate the fission products and prepare the remaining uranium and plutonium for recycling. The specific method used depends on the elements — U, Pu, Th — and their form — metal, oxide, carbide, etc. We describe the system that can be applied to typical slightly enriched uranium as oxide in zircaloy metal tubes as used in a PWR. The basis for this review is the experience over the period 1966–1972 at the Nuclear Fuel Services (NFS) plant at West Valley, N.Y. and the design of the reprocessing facility of Allied General Nuclear Services (AGNS) at Barnwell, S.C.

Upon receipt of the shipping cask the spent fuel is unloaded and stored for further radioactive cooling. The assemblies are then fed into a mechanical shear that cuts them into small pieces, e.g., 3 cm long, to expose the pellets. The pieces fall into baskets that are immersed in nitric acid which dissolves the uranium oxide, leaving the zircaloy "hulls." The aqueous solution resulting from this chop-leach process then proceeds to a solvent extraction (Purex) process. To understand the action of Purex, consider an experiment: Add oil to a vessel containing salt water and shake. When the mixture settles and the liquids separate, some salt has gone with the oil, i.e., it has been extracted from the water. In the Purex process the solvent is the organic compound tributyl phosphate (TBP) diluted with kerosene. Countercurrent flow of the aqueous and organic materials is maintained in a packed column as sketched in Fig. 21.3. Mechanical vibration of the plates allows liquids to pass through with maximum contact as the light component rises and the heavy component falls.

A flow diagram of the process of separating the various components of spent fuel is shown in Fig. 21.4. The amount of neptunium ($^{239}_{93}$Np, half-life 2.35 days) present is dependent on how fresh the spent fuel is. After a month of holding, the neptunium will be missing. The products then are in the form of nitrate solutions of uranium, plutonium, and a fission product stream that contains a diverse collection of chemical elements.

Consider the possible use of the uranium. It will have a U-235 content that is above or near that of natural U and thus is suitable for injection into an isotope separator for enrichment up to an appropriate fresh fuel level. The plutonium can be converted into the oxide form for shipment

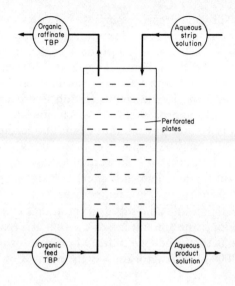

Fig. 21.3.   Solvent extraction by the Purex method.

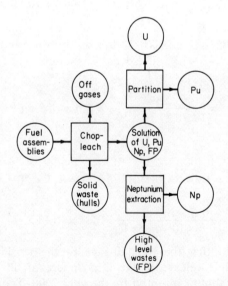

Fig. 21.4.   Simplified flow chart of nuclear fuel reprocessing.

to a fuel fabrication plant where it can be combined with uranium oxide to produce mixed oxide (MOX) as new fuel. New precautions in fabrication are required to protect workers against exposure to plutonium.

Returning to the subject of reprocessing, we note that special attention must be given to the release of certain radioactive gases. Among these are 12.346-yr tritium ($^3_1$H), which is a product of the occasional fission into three particles, 8.040-day iodine (I-131), and 10.701-yr krypton (Kr-85). The hazard associated with tritium is rather small because of the large MPC value, and the iodine concentration is greatly reduced by holding periods for fuel and process solutions. The long-lived krypton is a problem because it is a noble gas and thus cannot be reacted chemically for storage purposes. It may be disposed of in two ways — release to the atmosphere from tall stacks, with subsequent dilution by the atmosphere, or absorption of the chemically inert element in porous media such as charcoal operated at very low temperatures.

We can readily estimate the amount of krypton that must be handled. Since its half-life is long, the amount in $\frac{1}{3}$ of the fuel in a reactor operated for 3 yr is the same as that produced in the whole reactor in 1 yr. During the operation of a 3000 MWt reactor for 1 yr the number of fission events is

$$(3 \times 10^9 \text{ W})\left(3.2 \times 10^{10} \frac{\text{fissions}}{\text{W-sec}}\right)\left(3.16 \times 10^7 \frac{\text{sec}}{\text{yr}}\right) = 3 \times 10^{27} \text{ fissions/yr.}$$

Now the yield of Kr-85 in fission is 0.27%, and thus the number of atoms produced per year is 8.42×10^{24}. The decay constant is

$$\lambda = \frac{0.693}{t_H} = \frac{0.693}{(10.70)(3.16 \times 10^7)} = 2.0 \times 10^{-9} \text{ sec}^{-1}.$$

Then the activity is

$$A = N\lambda = (8.4 \times 10^{24})(2 \times 10^{-9}) = 1.7 \times 10^{16} \text{ d/sec}$$

or

$$A = \frac{1.7 \times 10^{16} \text{ d/sec}}{3.7 \times 10^{10} \text{ d/sec-Ci}} = 4.7 \times 10^5 \text{ Ci.}$$

It has been estimated that the total dosage to the world population resulting from complete release to the atmosphere of all Kr-85 from reactors would reach about 1 mrem/yr by the year 2000. Though this is a small figure, it is close enough to the 5 mrem/yr limit that steps will probably be taken to collect the gas and hold it for decay.

21.5 WASTE STORAGE AND DISPOSAL

After the uranium and plutonium are recovered, there remains a large volume of solution containing high-level fission products. Some of these such as strontium-90, cesium-137, and promethium-147 are valuable for industrial or medical applications, as will be described in Chapter 22. The total activity of such isotopes greatly exceeds the demand, however, and thus the bulk of the fission products have no presently known practical value. They must be stored or disposed of in such a way as to prevent exposure to the public.

Many ideas on what to do with these wastes have been advanced over the years since nuclear reactors started operating. For expediency, early byproducts of the plutonium production plants at Hanford, Washington were stored in solution in large underground tanks. Some consideration was given to the possibility of sealing concentrated wastes in concrete and dumping the containers in the ocean, but this concept was abandoned, since there was a chance of eventual erosion of the vessels, with the escape of long-lived radioactive materials.

A distinction must be made between *storage*, which implies that the material can be retrieved, and *disposal*, in which little or no future attention is required. A natural question arises—how long must wastes be retained? If absolute protection of the public from exposure is demanded, it is necessary to provide facilities that are perfect and truly permanent. The dominant isotope is Pu-239, which will be present in small amounts in wastes, even if the recovery is as high as 99.5%. The maximum permissible body burden for Pu-239 for occupational workers, as recommended by ICRP, is only 0.6 microgram, corresponding to an MPC in soluble form of $1 \times 10^{-4} \, \mu \text{Ci/cm}^3$ in water and $2 \times 10^{-12} \, \mu \text{Ci/cm}^3$ in air. It is fortunate that there are several obstacles to ingestion of plutonium by human beings, including low solubility, low uptake by plants, and small absorption by the body.

For storage that assures a high degree of safety, it is necessary to plan for very rigid control for many hundreds of years, during which the main fission products to consider are Sr-90 (28.8 yr) and Cs-137 (30.2 yr). In a period as long as 500 yr, which is 17.4 half-lives for strontium, the fraction that remains is still high, $2^{-17.4} = 6 \times 10^{-6}$.

Let us calculate the amount of Sr-90 produced per year from a 3000-MWt plant, yielding 3×10^{27} fissions. With a fission yield of 5.8%, and a decay constant $\lambda = 7.6 \times 10^{-10} \, \text{sec}^{-1}$, the activity initially is found to be 3.6 megacuries. It is reasonable to assume that the concentrated liquid wastes from fuel reprocessing amount to 100 gal per 10,000 MWd. In one year, the reactor energy is 1.1×10^6 MWd, and thus the liquid volume is $(11,000 \text{ gal})(3785 \text{ cm}^3/\text{gal}) = 4.16 \times 10^7 \text{ cm}^3$. Thus the activity of Sr-90 is about 0.09 Ci/cm^3. This is to be compared with the MPC for Sr-90 for drinking water, which is $3 \times 10^{-7} \mu$ Ci/cm^3. The ratio between these figures is 3×10^{11}, implying that the amount of leakage from a storage tank holding one year's production of Sr-90 to any source of public water must be restricted to essentially zero.

We now consider the main isotopes that contribute to the heating and radiation hazard for long times after fuel is removed from the reactor. Ample data are available on the fission products.* Table 21.2 shows the more important of these, selected on the basis of half-life and energy. When two isotopes are listed, the first, a parent, has the longer half-life, but the second, a daughter, also yields radiation.

Small but important amounts of heavy isotopes are also present in the waste. These are called the actinides, and are the result of successive neutron bombardment of uranium and plutonium isotopes. One sequence starts with radiative capture in U-235. By adding more neutrons with radioactive decay isotopes formed are U-236, U-237, Np-237, Np-238, and Pu-238, the latter with 87.7 year half-life.

Table 21.1. Principal Isotopes in Long-term Waste Storage

Isotope	Fission yield†	Half-life	Energy (MeV) Beta	Energy (MeV) Gamma
$^{90}_{38}$Sr $-^{90}_{39}$Y	0.0575	28.82 y, 64.06 h	1.129	——
$^{137}_{55}$Cs	0.0611	30.174 y	0.1744	0.6263
$^{144}_{58}$Ce $-^{144}_{59}$Pr	0.0545	284.5 d, 17.30 m	1.331	0.0602
$^{85}_{36}$Kr	0.0027	10.701 y	0.251	
$^{106}_{44}$Ru $-^{106}_{45}$Rh	0.0039	366.5 d, 29.9 s	1.456	0.1994
$^{129}_{53}$I	0.0066	1.57×10^7 y	0.062	0.040
$^{147}_{61}$Pm	0.0227	2.62344 y	0.063	——
$^{134}_{55}$Cs	‡	2.062 y	0.161	1.580
$^{3}_{1}$H	§	12.346 y	0.0057	——

†Directly and indirectly through parent isotopes.
‡Not a fission product but the result of neutron capture in $^{133}_{55}$Cs.
§Produced by ternary fission.

*For example, in the comprehensive reports by P. F. Rose and T. W. Burrows, *ENDF/B Fission Product Decay Data BNL-NCS-50545*, Vols. I, II, National Neutron Cross Section Center, Brookhaven National Laboratory, Upton, N.Y. (1976).

Another starts with U-238, giving rise to U-239, Np-239, Pu-239, Pu-240, Pu-241, Pu-242, Am-243, half-life 7370 years. The decay of fissile Pu-241, half-life 14.4 years, produces Am-241, half-life 432 years.

Many techniques for isolating high-level radioactive wastes have been proposed over the years since nuclear reactors started operating. Among the ideas on waste disposal that have been advanced are these:

(a) Allow canisters containing wastes to melt their way down into the Antarctic ice cap until they reach solid rock. This method would be suitable if the climate remained unchanged for many centuries.

(b) Send wastes into space — into orbit around the earth, or into orbit around the sun, or into deep space. This would entail excessive cost for the rockets to transport the material, and is objectionable on the grounds of contamination of space outside the earth.

(c) To irradiate the isotope with neutrons, converting them into short-lived or inert forms. Unfortunately the capture cross sections of the main offenders are so low that the fluxes must be prohibitively high. Only a fusion reactor would supply the required fluxes.

(d) To bury the wastes deep in the ground or in the sea bed. Several methods have been considered including dumping the wastes down a shaft leading to a prepared cavity, and allowing the fission product heat to fuse the rock and waste together.

In the favored isolation method, Fig. 21.5, wastes are dried and mixed with a glass to form a solid that is resistant to chemical attack. The material is placed in metal cylinders about 30 cm in diameter and 300 cm in length. The resulting canisters are then deposited in holes drilled in the floor of a tunnel cut deep in a salt bed. The heat of the fission products causes the salt to melt around the canisters and seal them in place. This technique has been studied thoroughly and appears to be suitable for very long-term storage, in areas where it can be demonstrated that the salt has not been nor is likely to be accessible to water.

In the event that spent fuel is not processed, the fuel assemblies would first be stored in water pools similar to those at reactor plants and subsequently would be sealed in containers and treated similarly to the solidified waste canisters.

Considerable research on waste isolation has been done in Sweden. Canisters composed of titanium and lead are proposed for vitrified wastes, with canisters of copper for spent fuel. Also considered is synthetic corundum, an extremely hard gem material. A multi-barrier approach is planned to achieve the necessary thousands of years of integrity.

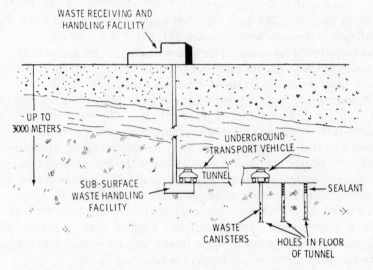

WASTE RECEIVING AND
HANDLING FACILITY

- UP TO
3000 METERS

UNDERGROUND
TRANSPORT VEHICLE

TUNNEL

SUB-SURFACE
WASTE HANDLING
FACILITY

SEALANT

WASTE
CANISTERS

HOLES IN FLOOR
OF TUNNEL

Fig. 21.5.   Nuclear waste isolation by geologic emplacement.

21.6 SUMMARY

Fission products constitute a waste from the operation of nuclear power plants. Although their volume is small, their radioactivity is extremely high, requiring that great care be exercised in the operations of fuel transfer, storage for decay, transportation, and reprocessing. In the U.S. this last step is not used and spent fuel is stored in water pools. If chemical processing is employed, uranium, plutonium, and fission products are separated. Because of their hazard, waste radioisotopes must be isolated for very long periods. Various methods for storage and disposal of wastes have been suggested but underground deposition seems preferable.

21.7 PROBLEMS

21.1. A batch of radioactive waste from a processing plant contains the following isotopes:

Isotope	Half-life	Fission yield, %
I-131	8.0 days	2.9
Ce-141	33 days	6
Ce-144	284 days	6.1
Cs-137	30.2 yr	5.9
I-129	1.57×10^7 yr	1

Letting the initial activity at $t = 0$ be proportional to λ and the fission yield, plot on semilog paper the activity of each for times ranging from 0 to 100 yr. Form the total and identify which isotope dominates at various times.

21.2. Traces of plutonium remain in certain waste solutions. If the initial concentration of Pu-239 in water were 100 parts per million (μg/g), find how much of the water would have to be evaporated to make the solution critical, neglecting neutron leakage as if the container were very large. Note: for H, $\sigma_a = 0.332$; for Pu, $\sigma_f = 742$, $\sigma_a = 1013$, $\nu = 2.87$.

21.3. If the maximum permissible concentration of Kr-85 in air is $1.5 \times 10^{-9} \, \mu$Ci/cm^3, and the yearly reactor production rate is 5×10^5 Ci, what is a safe diluent air volume flow rate (in cm^3/sec and ft^3/min) at the exit of the stack? Discuss the implications of these numbers in terms of protection of the public.

21.4. Calculate the decay heat from a single fuel assembly of the total of 189 in a 3000-MWt reactor at one day after shutdown of the reactor. How much longer is required for the heat generation rate to go down an additional factor of 2?

21.5. From the data given in Fig. 21.1 and the text:
(a) Deduce the percentages of the total power obtained from each of these isotopes: U-235, U-238, Pu-239, and Pu-241.
(b) Calculate the yearly weight of fission products in the spent fuel removed, assuming fifty-nine assemblies removed per year. Estimate the effective power level and the percentage of the time the 3000-MWt reactor was operated.

21.6. Assume that high-level wastes should be secured for a time sufficient for decay to reduce the concentrations by a factor of 10^{10}. How long is this in years for strontium-90? For cesium-137? For plutonium-239?

22

Beneficial Uses of Isotopes

Many important economic and social benefits are derived from the use of isotopes and radiation. The discoveries of modern nuclear physics have led to new ways to observe and measure physical, chemical, and biological processes, providing the strengthened understanding so necessary for man's survival and progress. The ability to isolate and identify isotopes gives additional versatility, supplementing techniques involving electrical, optical, and mechanical devices.

Special isotopes of an element are distinguishable and thus traceable by virtue of their unique weight or their radioactivity, while behaving chemically as the other isotopes of the element. Thus it is possible to measure amounts of the element or its compounds and trace movement and reactions.

When one considers the thousands of stable and radioactive isotopes available and the many fields of science and technology that require knowledge of process details, it is clear that a catalog of possible isotope uses would be voluminous. We shall be able here only to compare the merits of stable and radioactive species, to describe some of the special techniques, and to mention a few interesting or important applications of isotopes.

22.1 STABLE AND RADIOACTIVE ISOTOPES

Stable isotopes, as their name suggests, do not undergo radioactive decay. Most of the isotopes found in nature are in this category and appear in the element as a mixture. The principal methods of separation

230

according to isotopic mass are electromagnetic, as in the large-scale mass spectrograph; and thermal-mechanical, as in the distillation or gaseous diffusion processes. Important examples are isotopes of elements involved in biological processes, e.g., deuterium and oxygen-18. The main advantages of stable isotopes are the absence of radiation effects in the specimens under study, the availability of an isotope of a chemical for which a radioactive species would not be suitable, and freedom from concern with speed in making measurements, since the isotope does not decay in time. Their disadvantage is the difficulty of detection.

Radioactive isotopes, or radioisotopes, are available with a great variety of half-lives, types of radiation, and energy. They come from three main sources—charged particle reactions in an accelerator, neutron bombardment in a reactor, and separated fission products. For example, the stable isotope of iodine is I-127; bombardment of tellurium-124 by deuterons yields a neutron and I-125 of half-life 60 days; absorption of neutrons in I-127 gives I-128, half-life 25 min; one of the iodine isotopes from fission is I-131, half-life 8 days. The main advantages of using radioisotopes are ease of detection of their presence through the emanations, and the uniqueness of the identifying half-lives and radiation properties. We shall now describe several special methods involving radioisotopes and illustrate their use.

22.2 TRACER TECHNIQUES

The tracer method consists of the introduction of a small amount of an isotope and the observation of its progress as time goes on. For instance, the best way to apply fertilizer containing phosphorus to a plant may be found by including minute amounts of the radioisotope phosphorus-32, half-life 14.28 days, emitting 1.7 MeV beta particles. Measurements of the radiation at various times and locations in the plant by a detector or photographic film provides accurate information on the rate of phosphorus intake and deposition. Similarly, circulation of blood in the human body can be traced by the injection of a harmless solution of radioactive sodium, Na-24, 15.03-hr half-life. For purposes of medical diagnosis, it is desirable to administer enough radioactive material to provide the needed data, but not so much that the patient is harmed.

The flow rate of many materials can be found by watching the passage of admixed radioisotopes. The concept is the same for flows as diverse as blood in the body, oil in a pipeline, or pollution discharged into a river. As sketched in Fig. 22.1, a small amount of radioactive material is injected at

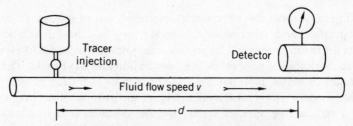

Fig. 22.1. Tracer measurement of flow rate.

a point, it is carried along by the stream, and its passage at a distance d away at time t is noted. In the simplest situation, the average fluid speed is d/t. It is clear that the half-life of the tracer must be long enough for detectable amounts to be present at the point of observation but not so long that the fluid remains contaminated by radioactive material.

In many tracer measurements for biological or engineering purposes, the effect of removal of the isotope by other means besides radioactive decay must be considered. Suppose, as in Fig. 22.2, that liquid flows in and out of a tank of volume V (cm^3) at a rate v (cm^3/sec). A tracer of initial amount N_0 atoms is injected and assumed to be uniformly mixed with the contents. Each second, the fraction of fluid (and isotope) removed from the tank is v/V, which serves as a flow decay constant λ_f for the isotope. If radioactive decay were small, the counting rate from a detector would decrease with time as $e^{-\lambda_f t}$. From this trend, one can deduce either the speed of flow or volume of fluid, if the other quantity is known. If both radioactive decay and flow decay occur, the exponential formula may also be used but with the effective decay constant $\lambda_e = \lambda + \lambda_f$. The composite effective half-life then can be found from the

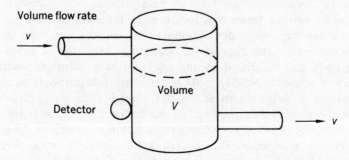

Fig. 22.2. Flow decay.

relation

$$\frac{1}{(t_H)_e} = \frac{1}{t_H} + \frac{1}{(t_H)_f}.$$

This logic applies equally well to the injection of a radioactive substance to an organism. The "biological half-life" takes the place of the flow half-life.

22.3   NEUTRON ACTIVATION ANALYSIS

This is an analytical method that will reveal the presence and amount of minute impurities. A sample of material that may contain traces of a certain element is irradiated with neutrons, as in a reactor. The gamma rays emitted by the product radioisotope have unique energies and relative intensities, in analogy to spectral lines from a luminous gas. Measurements and interpretation of the gamma ray spectra, using data from standard samples for comparison, provide information on the amount of the original impurity.

Let us consider a practical example. Reactor design engineers may be concerned with the possibility that some stainless steel to be used in moving parts in a reactor contains traces of cobalt, which would yield undesirable long-lived activity if exposed to neutrons. To check on this possibility, a small sample of the stainless steel is irradiated in a test reactor to produce Co-60, and gamma radiation from the Co-60 is compared with that of a piece known to contain the radioactive isotope. The "unknown" is placed on a Pb-shielded large-volume lithium-drifted germanium Ge(Li) detector used in gamma-ray spectroscopy as noted in Section 11.4. Gamma rays from the decay of the 5.27-yr Co-60 give rise to electrons by photoelectric absorption, Compton scattering, and pair production. The electrons then produce electrical signals in the detector in approximate proportion to the energy of the gamma rays. If all of the pulses produced by gamma rays of a single energy were equal in height, the observed counting rate would consist of two perfectly sharp peaks at energy 1.17 MeV and 1.33 MeV. A variety of effects causes the response to be broadened somewhat as shown in Fig. 22.3. The location of the peaks clearly shows the presence of the isotope Co-60 and the heights tell how much of the isotope is present in the sample. Modern electronic circuits can process a large amount of data at one time. The multichannel analyzer accepts counts due to photons of all energy and displays the whole spectrum graphically.

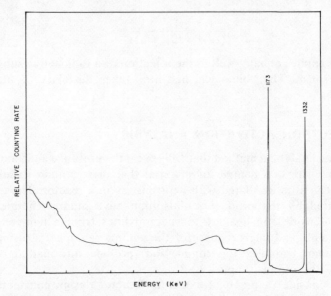

Fig. 22.3.   Analysis of gamma rays from cobalt-60. (Courtesy of Jack N. Weaver of North Carolina State University.)

When neutron activation analysis is applied to a mixture of materials, it is necessary after irradiation to allow time to elapse for the decay of certain isotopes whose radiation would "compete" with that of the isotope of interest. In some cases, prior chemical separation is required to eliminate interfering isotope effects.

The activation analysis method is of particular value for the identification of chemical elements that have an isotope of high neutron absorption cross section, and for which the products yield a suitable radiation type and energy. Not all elements meet these specifications, of course, which means that activation analysis supplements other techniques. For example, neutron absorption in the naturally occurring isotopes of carbon, hydrogen, oxygen, and nitrogen produces stable isotopes. This is fortunate, however, in that organic materials including biological tissue are composed of those very elements, and the absence of competing radiation makes the measurement of trace contaminants easier. The sensitivity of activation analysis is remarkably high for many elements, it being possible to detect quantities as low as a millionth of a gram in 76 elements, a billionth of a gram in 53, or even as low as a trillionth in 11.

22.4 APPLICATIONS OF ACTIVATION ANALYSIS

A few of the many practical uses of the method are cited.

(a) Textile Manufacturing

In the production of synthetic fibers, certain chemicals such as fluorine are applied to improve textile characteristics, such as the ability to repel water or stains. Activation analysis is used to check on inferior imitations, by comparison of the content of fluorine or other deliberately added trace elements.

(b) Petroleum Processing

The "cracking" process for refining oil involves an expensive catalyst that is easily poisoned by small amounts of vanadium, which is a natural constituent of crude oil. Activation analysis provides a means for verifying the effectiveness of the initial distillation of the oil.

(c) Crime Investigation

The process of connecting a suspect with a crime involves physical evidence that often can be accurately obtained by activation analysis. Examples are: the comparison of paint flakes found at the scene of an automobile accident with paint from a hit-and-run driver's car; the determination of the geographical source of drugs such as opium by comparison of trace element content with that of soils in which the poppy plant can be grown; the verification of the source of copper telephone wire believed to be stolen, by taking account of variations in the chemical content of wire from different manufacturers; measurement of the amounts of barium or antimony on the hands of suspects or victims in gunshot cases where it is not clear whether murder or suicide is involved; the identification of the origin of bullets by their unique antimony content; and tests for poison in a victim's body. The classic example of the latter is the verification of the hypothesis that Napoleon was poisoned, by use of activation analysis measurements of the content of arsenic in samples of his hair.

(d) Authentication of Paintings

The probable age of a work of art can be found by testing a small speck of paint. Over the centuries, the proportions of elements such as chromium and zinc used in paint have changed, and forgeries of the work of old masters can thus be detected.

(e) Diagnosis of Disease

Promising medical applications still under investigation include the measurement of sodium content in children's fingernails in the diagnosis of cystic fibrosis, and the accurate measurement of normal and abnormal content of some 50 trace elements in the blood, as indicators of specific diseases.

(f) Pesticide Investigation

The amounts of residues of pesticides such as DDT or methyl bromide in crops, foods, and animals are found by analysis of the bromine and chlorine content.

(g) Mercury in the Environment

The heavy element mercury is a serious poison for animals and human beings even at low concentrations. It appears in rivers as a result of discharge of certain manufacturing process wastes. By the use of activation analysis, which can measure down to around $\frac{1}{20}$ of a part per million of mercury, the amount of mercury contamination in water or tissues of fish or land animals can be determined, helping establish the chain by which it is transferred to biological species.

22.5 GAGING

Some physical properties of materials are difficult to ascertain by ordinary methods, but can be measured readily by observing how radiation interacts with the substance. For example, the thickness of a thin layer of plastic or paper can be found by measuring the number of beta particles from a radioactive source that are transmitted. The separated fission product isotopes Sr-90 (28.82 yr, 0.546 MeV β particle) and Cs-137 (30.17 yr, 0.512 MeV β particle) are widely used for such gaging.

The density of a liquid flowing in a pipe can be measured externally by detection of the gamma rays that pass through the substance. The liquid in the pipe acts as a "shield" for the radiation, with attenuation dependent on macroscopic cross section and thus particle number density, as discussed in Chapter 19.

The level of liquid in an opaque container can be measured readily without the need for sight glasses or electrical contacts. A detector outside the vessel measures the radiation from a radioactive source mounted on a float in the liquid.

The moisture content of soils can be estimated by study of the neutrons slowed by hydrogen. In the neutron moisture gage, a source consisting of a mixture of an alpha particle emitter, e.g., Pu-239, and beryllium Be-9 provides fast neutrons by the (α, n) reaction. The flux of thermal neutrons measured by a BF_3 counter provides data on the water content.

Several nuclear techniques are employed in the petroleum industry. In the drilling of wells, the "logging" process involves the study of geological features. One method consists of measurement of natural gamma radiation. When the detector is moved from a region of natural radioactive rock to one containing oil or other liquid, the signal is reduced. The neutron moisture gage is also adapted to determine the presence of oil, which contains hydrogen. Neutron activation analysis of chemical composition is performed by lowering a neutron source and a gamma ray detector into the well.

22.6 DATING

There would appear to be no relation between nuclear energy and the humanities such as history, archeology, and anthropology. There are, however, several interesting examples in which nuclear methods establish dates of events. The carbon dating technique is being used regularly to determine the age of ancient artifacts. The technique is based on the fact that carbon-14 is and has been produced by cosmic rays in the atmosphere (a neutron reaction with nitrogen). Plants take up CO_2 and deposit C-14, while animals eat the plants. At the death of either, the supply of radiocarbon obviously stops and that present decays, with half-life 5730 yr. By measurement of the radioactivity, the age within about 50 yr can be found. This method was used to determine the age of the Dead Sea Scrolls, as about 2000 yr, making measurements on the linen made from flax; to date the documents at Stonehenge in England, using pieces of charcoal; and to verify that prehistoric peoples lived in the United States, as long ago as 9000 yr, from the C-14 content of rope sandals discovered in an Oregon cave.

Even greater accuracy in dating biological artifacts can be obtained by direct detection of carbon-14 atoms. Molecular ions formed from $^{14}_{6}C$ are accelerated in electric and magnetic fields and then slowed by passage through thin layers of material. This sorting process can measure 3 atoms of $_6^{14}C$ out of 10^{16} atoms of $_6^{12}C$.

The age of minerals in the earth, in meteorites, or on the moon can be obtained by a comparison of their uranium and lead contents. The method is based on the fact that Pb-206 is the final product of the decay chain starting with U-238, half-life 4.468×10^9 yr. Thus the number of lead atoms now present is equal to the loss in uranium atoms, i.e.,

$$N_{Pb} = (N_U)_0 - N_U,$$

where

$$N_U = (N_U)_0 e^{-\lambda t}.$$

Elimination of the original number of uranium atoms $(N_U)_0$ from these two formulas gives a relation between time and the ratio N_{Pb}/N_U. The latest value of the age of the earth obtained by this method is 4.55 billion years.

For the determination of ages ranging from 50,000 to a few million years, an argon method can be employed. It is based on the fact that the potassium isotope K-40 (half-life 1.277×10^9 yr) crystallizes in materials of volcanic origin and decays into the stable argon isotope Ar-40. The technique is of particular interest in attempting to establish the date of the first appearance of man.

22.7 APPLICATIONS OF WASTES

The very isotopes that constitute a hazardous waste from nuclear energy production can provide protection and increased safety. One example is the use of 10.7-yr Kr-85, noted in Chapter 21 to be a copious byproduct of fuel reprocessing. The isotope can be the active component in self-luminous light sources for airport runways and coal mines. The sources consist of a sealed capsule of Kr-85 gas which is in contact with a phosphor that is excited by low-energy electrons. The color depends on the phosphor used, the brightness on the amount of isotope. The sources resemble an auto headlight with its bulb replaced by the krypton capsule. Their virtue is long life, independence of power source, and indifference to weather conditions.

Security against intruders can be furthered by the use of Kr-85 "beam breakers" in which a thin pencil of low-energy gamma rays is attenuated by the presence of a human body and an alarm is set off. Such devices may have use in providing safeguards against diversion of nuclear materials.

One type of electronic smoke detector for use in fire protection employs 2 μCi of the actinide isotope americium-241, an alpha emitter (5.5 MeV) with half-life 432 years (see also Section 21.5). The radiation provides ionization of the air between two electrodes, and changes in conductivity due to smoke cause the circuit to alarm.

22.8 SUMMARY

Stable and radioactive isotopes have great utility for measurements of properties and processes in many fields. The tracer technique provides information on flow of fluids, neutron activation analysis tells the amounts of minute impurities in a great variety of applications, and isotope gages measure thickness, density, and moisture content. Accurate dating of ancient specimens is made possible by measurements of C-14, of argon from decay of K-40, and of the ratio of lead to uranium. Certain fission product isotopes that are nuclear wastes can be used in the interests of safety.

22.9 PROBLEMS

22.1. A radioisotope is to be selected to provide the signal for arrival of a new grade of oil in an 800-km-long pipe line, in which the fluid speed is 1.5 m/sec. Some of the candidates are:

Isotope	Half-life	Particle, energy (Mev)
Na-24	15.030 h	β, 1.389; γ, 1.369, 2.754
S-35	87.39 d	β, 0.167
Co-60	5.2719 y	β, 0.318; γ, 1.173, 1.332
Fe-59	44.56 d	β, 0.273, 0.475; γ, 1.099, 1.292

Which would you pick? On what basis did you eliminate the others?

22.2. The radioisotope F-18, half-life 110 min, is used for tumor diagnosis. It is produced by bombarding lithium carbonate (Li_2CO_3) with neutrons, using tritium as an intermediate particle. Deduce the two nuclear reactions.

22.3. The range of beta particles of energy 0.53 MeV in metals is 170 mg/cm^2. What is the maximum thickness of aluminum sheet, density 2.7 g/cm^3, that would be practical to measure with a Sr-90 or Cs-137 gage?

22.4. The amount of environmental pollution by mercury is to be measured using neutron activation analysis. Neutron absorption in the mercury isotope Hg-196, present with 0.15% abundance, activation cross section 3×10^3 barns, produces the radioactive species Hg-197, half-life 64.14 hr. The smallest activity for which the resulting photons can be accurately analyzed in a river water sample is 10 d/sec. If a reactor neutron flux of 10^{12} cm^{-2}-sec^{-1} is available, how long an irradiation is required to be able to measure mercury contamination of 20 ppm (μg/g) in a 4 milliliter water test sample?

22.5. The ratio of numbers of atoms of lead and natural uranium in a certain moon rock is found to be 0.05. What is the age of the sample?

22.6. The activity of C-14 in a wooden figure found in a cave is only $\frac{3}{4}$ of today's value. Estimate the date the figure was carved.

22.7. Examine the possibility of adapting the uranium–lead dating analysis to the potassium–argon method. What would be the ratio of Ar-40 to K-40 if a deposit were 1 million years old?

22.8. The age of minerals containing rubidium can be found from the ratio of radioactive Rb-87 to its daughter Sr-87. Develop a formula relating this ratio to time.

22.9. It has been proposed to use radioactive krypton gas of 10.7-yr half-life in conjunction with film for detecting small flaws in materials. Discuss the concept, including possible techniques, advantages, and disadvantages.

22.10. A krypton isotope $^{81m}_{36}$Kr of half-life 13 seconds is prepared by charged particle bombardment. It gives off a gamma ray of 0.19 MeV energy. Discuss the application of the isotope to the diagnosis of emphysema and black-lung disease. Consider production, transportation, hazards, and other factors.

22.11. Tritium ($^3_1$H) has a physical half-life of 12.346 years but when taken into the human body as water it has a biological half-life of 12.0 days. Calculate the effective half-life of tritium for purposes of radiation exposure. Comment on the result.

23

Applications of Radiation

The nuclear methods just described emphasized the detection and measurement of radiation as a means to identify the emitters. We now turn to the beneficial effects of radiation from various sources—X-ray machines, charged particle accelerators, nuclear reactors, and radioisotope sources. The penetrating radiations are electromagnetic, electrons or other charged particles, and neutrons. Applications are cited in industry, medicine, agriculture, and space exploration, with the examples selected for importance and interest.

Appreciation is extended to William C. Remini of the Department of Energy for his suggestions on this chapter.

23.1 RADIOGRAPHY

The oldest and most familiar beneficial use of radiation is for medical diagnosis by X-rays. These consist of high-frequency electromagnetic radiation produced by electron bombardment of a heavy-metal target. As is well known, X-rays penetrate body tissue to different degrees dependent on material density, and shadows of bones and other dense material appear on the photographic film. The term radiography includes the investigation of internal composition of living organisms or objects in general, using X-rays, gamma rays, or neutrons.

For both medical and industrial use, the isotope Co-60, produced from stable Co-59 by neutron absorption, is an important alternate to the X-ray tube. Co-60 provides gamma radiation of energies 1.17 and 1.33 MeV, which are especially useful for examination of flaws in metals. Internal

241

cracks, defects in welds, and nonmetallic inclusions are revealed by scanning with a cobalt radiographic unit. Advantages include small size and portability, and freedom from the requirement of an electrical power supply. The half-life of 5.27 yr permits use of the cobalt source for a long time without refueling. On the other hand, the energy of rays is fixed and the intensity cannot be varied, as is possible with the X-ray machine. For radiography of thin specimens, the isotope iridium-192 is convenient. Its half-life is 74.2 days and the photon energies around 0.4 MeV.

Neutron radiography serves as a complement to gamma ray radiography where the materials are insensitive to photons but are rich in hydrogen, e.g., plastics and rubber. The neutron sources are typically antimony–beryllium, in which the gamma rays from Sb-124, half-life 60.4 days, cause a (γ, n) reaction in Be-9. A promising new source is californium-252, an artificial isotope (element 98) produced by successive neutron bombardment of plutonium in a reactor. Although Cf-252 decays most of the time (96.908%) by alpha particle emission, it undergoes spontaneous fission the rest of the time (3.092%). The corresponding half-lives for the two processes are 2.730 yr and 85.57 yr respectively. The fission gives around 3.5 neutrons on the average. An extremely small mass of the isotope thus serves as an abundant source of fast neutrons.

23.2 MEDICAL TREATMENT

A rapid growth in the use of radiation for medical therapy has taken place in recent years, with millions of treatments of patients administered annually. The radiation can come from teletherapy units, in which the source is at some distance from the patients, or from isotopes in sealed containers implanted in the body, or from injected or ingested solutions containing isotopes.

Doses of radiation are often found to be effective in the treatment of certain diseases such as cancer. Over the years, X-rays have traditionally been used, but it has been found that the penetrating Co-60 gamma rays permit higher doses to tissue deep in the body, with a minimum of skin reaction. The cobalt equipment requires no expensive electrical maintenance.

Considerable success in treatment of abnormal pituitary glands has been obtained by irradiations with charged particles from an accelerator, and slow neutron irradiation of tumors injected with boron solution has been found beneficial in some cases. Diseases such as leukemia and hyperthyroidism respond to treatment by radiation from radioisotopes of phosphorus and iodine, respectively.

23.3   PATHOGEN REDUCTION

Gamma radiation of energy 0.662 MeV as emitted by cesium-137 (half-life 30.174 yr) has been shown to be effective in reducing pathogens (viruses, bacteria, and parasites) and thus can be used for such tasks as sterilizing medical supplies. The prevention of infections in surgery requires the use of sutures that have been sterilized. Conventional inefficient batch processing with chemicals and heat has been largely replaced by mass-production methods involving the bombardment of packaged sutures by radiation.

Another promising application is the elimination of pathogens from dried sewage sludge. A dose of 30 kilorad has been shown to effect a 90% reduction in bacteria. Treated wastes can be used safely as a fertilizer or animal food supplement. Such applications are important in terms of both environmental protection and recycling of materials for energy conservation.

23.4   FIBER IMPROVEMENT

Various properties of polymers such as polyethylene can be changed by electron or gamma ray irradiation. The original material consists of very long parallel chains of molecules, and the radiation causes the chains to be connected, a process called cross-linking. Irradiated polyethylene has better resistance to heat and serves as excellent electrical insulating coating for wires. Fabrics can be made soil-resistant by a process of radiation bonding of a suitable polymer to a fiber base.

23.5   SYNTHESIS OF CHEMICALS

Certain chemical reactions can be initiated using high-energy gamma radiation. Many of these reactions are feasible in a laboratory, but relatively few processes can be made economical. An exception is the production of ethyl bromide (CH_3CH_2Br), a volatile organic liquid used as an intermediate compound in the synthesis of organic materials. Gamma radiation from a Co-60 source has the effect of a catalyst in the combination of hydrogen bromide (HBr) and ethylene (CH_2CH_2). Gammas are found to be superior as catalysts to chemicals, or application of ultraviolet light, or electron bombardment. Millions of pounds of ethyl bromide are produced annually by this unique process.

The commercially important chemical polyethylene is also produced by cobalt gamma ray bombardment of ethylene.

23.6 WOOD PLASTIC PROCESSING

There is a large commercial demand for a new type of wood flooring produced by gamma irradiation. Wood is soaked with a plastic and passed through a beam of gamma rays, which changes the molecular structure of the plastic and leaves a surface that cannot be scratched or burned. The appearance of the product is unchanged, but the material is extremely durable, making the wood especially useful for public areas such as lobbies of airport terminals. The extra cost of processing is justified by the long useful life of the material, and millions of square feet are manufactured each year. Similar techniques of irradiation are used in the preparation of architectural tiles with much improved wear and strength characteristics.

23.7 RADIATION PRESERVATION OF FOOD

As early as 1953, studies were under way on the feasibility of large-scale processing of food by irradiation with gamma rays. It has been shown repeatedly that dramatic improvements in the shelf life of foods are effected, of the order of months, and there is no evidence that radioactivity is induced by gamma ray bombardment. There are two unresolved problems, however—in some foods, subtle changes in taste are noted, some of which are disagreeable, and there is concern that the radiation may induce chemical changes that render the food unsafe.

Among examples of produce that have been tested are potatoes (to inhibit sprouting), bacon, wheat (for disinfestation), strawberries (to prevent decay and rot), and fish. Research is under way on the organism that produces botulism in fish products. A jurisdictional problem involving federal agencies—the Food and Drug Administration, the Atomic Energy Commission, and the Army—unfortunately slowed progress in this important area and future applications are uncertain.

23.8 CROP MUTATIONS

The science of crop breeding involves the selection of unusual plants and crossing them to obtain permanent and reproducible hybrids with desired properties. Crops with high yield, resistance to disease and adaptability to new environments have been obtained by such genetic studies. However, the process can be accelerated by the application of radiation, such as charged particles, X-rays, gamma rays, and neutrons. Desirable mutations are induced by irradiation of seeds or by the removal of cuttings from irradiated trees. Success in obtaining new strains has been achieved in beans, oats, barley, peanuts, and many types of ornamental flowers and plants.

The improvement of food production by crop mutations is of especial importance in the problem of an expanding world population, in terms of both a higher yield of crops and a higher nutritional value. For example, a new strain of mutant rice has been developed that contains twice as much protein as conventional varieties.

23.9 INSECT CONTROL

To eradicate certain insect pests, the so-called sterile male technique has been successfully applied. The procedure for eradication consisted of laboratory breeding of large numbers of male insects, sterilizing them with gamma radiation, and releasing them for mating in the infested area. Competition of the sterile males with normal males causes a rapid reduction in population over a two-year period. The most dramatic example is the elimination in the South and Southwest of the screw worm fly, a pest that had caused millions of dollars of damage to livestock. The fly lays eggs in wounds and the larva kills the animal.

The method also has promise for eradicating the tsetse fly, the carrier of the disease sleeping sickness, which is very prevalent in Africa. Many millions of acres of land are now uninhabitable because of the presence of the insect. The main problem to be solved is the technique for rearing and sterilizing the flies in adequate numbers.

23.10 ISOTOPIC POWER GENERATORS

The long half-life of certain radioisotopes make them very suitable in the construction of light, compact, and reliable power sources, especially for remote locations. One of the most important of such isotopes is Pu-238, half-life 86.4 yr, which emits alpha particles of 5.5 MeV energy. The isotope $^{238}_{94}$Pu is produced by reactor neutron irradiation of the almost-stable isotope $^{237}_{93}$Np (2.14×10^6-yr half-life). The latter is a decay product of $^{237}_{92}$U, a 6.75-day beta emitter that arises from neutron capture in $^{236}_{92}$U or by (n, 2n) and (γ, n) reactions with $^{238}_{92}$U. The high-energy alpha particles and the relatively short half-life of Pu-238 give the isotope the favorable power to weight ratio 0.57 W/g and a high specific activity of 17 Ci/g.

A number of space missions have used nuclear electric generators. Examples are Apollo for lunar science experiments (74 W), Pioneer for Jupiter–Saturn exploration (80 W), Viking for landing on Mars (85 W) and Voyager for trips to Jupiter, Saturn, and Uranus (475 W). Each of these missions used a radioisotope generator (RTG).

Typical of the RTGs is the one sent to the moon in the Apollo-12 mission. It powered a group of scientific instruments called ALSEP (Apollo Lunar Surface Experimental Package) which measured magnetic fields, dust, the solar wind, ions, and earthquake activity. The generator is shown schematically in Fig.23.1. Lead-telluride thermoelectric couples are placed between the PuO_2 and the beryllium case. Data on the generator are:

> System weight 20 kg.
> Pu-238 weight 2.6 kg.
> Activity 44,500 Ci.
> Capsule temperature 732°C.
> Thermal power 1480 W.
> Electrical power 74 W.
> Electrical voltage 16 V.
> Operating range −173°C to 121°C.

Research and development continues in a program called Kilowatt Isotope Power System (KIPS). The purpose is to achieve a power plant for sattelites of the 1980s and beyond. KIPS is to produce 0.5 to 2 kW of electric power by means of Pu-238 oxide, using an organic fluid Rankine cycle. The goal is a unit that will operate for 2500 days unattended.

A very promising medical spinoff of the development of the isotopic generator developed for the space program is the heart pacemaker, which provides small electrical impulses to regulate heartbeat. Pacemakers of a few hundred microwatts, powered by small quantities of Pu-238, will last for many years and are preferable to those powered by batteries, requiring frequent replacement by surgical operation. Such long life makes the isotopic source attractive for brain pacemakers, which stop epileptic seizures.

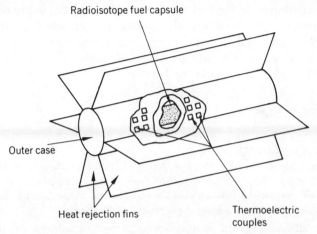

Radioisotope fuel capsule

Outer case

Heat rejection fins

Thermoelectric couples

Fig. 23.1.   Isotopic electrical power generator. (SNAP-27 used in Apollo-12 mission.)

Success with power sources for space applications prompted a program to develop a nuclear-powered artificial heart. Bench tests have been performed on a unit designed to be implanted in a calf. The basic components are (a) the 32 W Pu-238 isotopic power source, (b) a Stirling engine (a closed-cycle piston engine) as thermal converter, using argon as working fluid, (c) a mechanical blood pump, and (d) artificial plastic ventricles. Power of up to 3 W is available to circulate blood. Much disagreement developed as to the wisdom of using plutonium in the human body, and the program has been phased out, one would hope only temporarily, since heart disease is the No. 1 killer in the U.S.

Other isotopes that can be used for remote unattended heat sources are Sr-90 in the form of SrF_2 and Cs-137 as CsCl. When oil-fired units are not possible because of problems in fuel delivery or operability, an isotopic source is very practical. They cannot be justified on the basis of cost, however, since the expense of separating the isotopes from other fission products is equivalent to about $15/gal of fuel oil.

23.11   SUMMARY

Many applications of penetrating radiations are found in industry, medicine, agriculture, and space exploration. Radiography uses X-rays, Co-60 gamma rays, and neutrons to inspect the body or bulk materials.

Various radiations are applied for therapeutic benefit, improvement of fibers, synthesis of chemicals, the production of wood plastic, food preservation, beneficial crop mutations, and elimination of certain insect pests. Isotopic generators provide heat and electrical energy for space experiments and are available for heart pacemakers.

23.12 PROBLEMS

23.1. Using half-life relationships as given in Section 22.2, calculate the effective half-life of californium-252.

23.2. The half-life of Cf-252 is 85.57 yr. Assuming that it releases 3.5 neutrons per fission, how much of the isotope in micrograms is needed to provide a source of strength of 10^7 neutrons/sec? What would be the diameter of the source in the form of a sphere if the Cf-252 had a density as pure metal of 20 g/cm^3?

23.3. Three different isotopic sources are to be used in radiography of steel in ships as follows:

Isotope	Half-life	Gamma energy (MeV)
Co-60	5.27 yr	1.25 (ave.)
Ir-192	74.2 days	0.4 (ave.)
Cs-137	30.2 yr	0.66

Which isotope would be best for insertion in pipes of small diameter and wall thickness? For finding flaws in large castings? For more permanent installations? Explain.

23.4. A cobalt source is to be selected for irradiation of potatoes to inhibit sprouting. What strength in curies is needed to process 250,000 kg of potatoes per day, providing a dose of 10,000 rad? Note that two gammas totaling around 2.5 MeV energy are emitted by Co-60. What amount of isotopic power is involved? Discuss the practicality of absorbing all of the gamma energy in the potatoes.

23.5. (a) Verify that Pu-238, half-life 87.7 yr, alpha energy 5.5 MeV, yields an activity of 17 Ci/g and a power of 0.56 W/g.
(b) How much plutonium would be needed for a 200 μW heart pacemaker?

24

Nuclear Explosives

The intent of the present book is to emphasize the beneficial applications of nuclear energy, recognizing of course that the threat of warfare with nuclear weapons exists. However, since fissile material such as U-235 and Pu-239 can be used for either power or destruction, certain safeguards must be exercised in any civilian nuclear power program to avoid diversion of strategic materials for illegal purposes. Some understanding of what is involved in making bombs is essential to realistic decisions about controls on peaceful nuclear power. In this chapter we shall describe the nuclear explosion on an unclassified basis, review the means by which proliferation of nuclear weapons is discouraged, and mention some of the potential applications of explosives for peaceful purposes.

24.1 THE NUCLEAR EXPLOSION

Security of information on the detailed construction of nuclear weapons has been maintained, and only a qualitative description is available to the public. We shall draw on unclassified sources (see Appendix) for the following discussion.

First, we note that two types of devices have been used: (a) the fission explosive ("atom bomb") using plutonium or highly enriched uranium and (b) the fusion or thermonuclear explosive ("hydrogen bomb"). The reactions described in earlier chapters are involved. Next, it is possible to create an explosive fission chain reaction by two different procedures: (a) bringing

249

together rapidly two chunks of fissile material to achieve a supercritical mass — the so-called "gun" technique and (b) compressing a sphere of uranium or plutonium by application of concentrated high explosives, the "implosion" method. Pressures of more than 10 million pounds per square inch are required. Figure 24.1 shows these devices schematically. In either case, the large reactivity causes a rapid increase in power and the accumulated energy blows the material apart, a process labeled "disassembly." There are competing effects as compression reduces the core radius — an increase in ratio of surface to volume results in larger neutron leakage, but the decrease in mean free path reduces leakage. The latter effect dominates, giving a net positive increase of multiplication. (Also see Problem 12.3.) An unreflected plutonium assembly has a considerably lower critical mass, 16 kg, than U-235, 50 kg. By adding a 1-inch layer of natural U, the mass drops to 10 kg. The critical mass of uranium with reflector varies rapidly with the U-235 enrichment, as shown in Table 24.1. It is noted that the total mass of a

Table 24.1. Critical Masses of U-235 and U vs. Enrichment.

% U-235	U-235 (kg)	U (kg)
100	15	15
50	25	50
20	50	250
10	130	1300

device composed of less than 10 percent U-235 is impractically large. Thermonuclear explosives involve a central core of fissile material surrounded by layers of lithium-6 deuteride and U-238, as sketched in Fig. 24.2. The fissile material explodes, producing heat and neutrons, which convert some of the lithium-6 to tritium by the reaction discussed in Section 16.1,

$$^{6}_{3}\text{Li} + {}^{1}_{0}\text{n} \rightarrow {}^{3}_{1}\text{H} + {}^{4}_{2}\text{He} + 4.8\,\text{MeV}.$$

The heat causes D–D and D–T reactions to occur, which produce fast neutrons that cause fission in the U-238, enhancing the energy and radiation produced by the device.

The energy release or yield of a fission weapon depends on the degree of supercriticality that is reached before disassembly begins. The isotope Pu-240, formed by neutron capture in Pu-239 in a reactor, has an important effect on the reaction, since this isotope undergoes spontaneous fission. If too much Pu-240 is present, the neutrons released cause premature detonation and inefficient use of the fissile material. "Weapons grade" Pu contains much less Pu-240 than does "reactor grade" Pu.

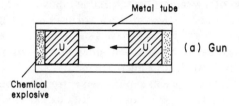

(a) Gun

Chemical
explosive

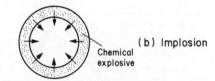

(b) Implosion

Chemical
explosive

Fig. 24.1. Fission-based explosive devices.

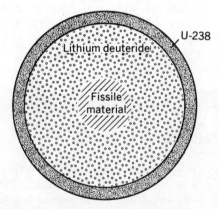

Fig. 24.2. Thermonuclear weapon. (From *The Great Test-Ban Debate* by Herbert F. York. Copyright © November 1972 by Scientific American, Inc. All rights reserved.)

Nuclear explosives release their energy in several ways. First is the blast effect, in which a shock wave moves outward in air, water, or rock, depending on where the event occurs. Second is the thermal radiation from the heated surrounding material, at temperatures typically 6000°C. Finally, there is the nuclear radiation, consisting mainly of neutrons and gamma rays. The fractions of the energy that go into these three modes are respectively 50, 35, and 15.

The energy yield of a weapon is measured in equivalent tons of chemical explosive. By convention, 1 ton of TNT corresponds to 10^9 calories of energy. The first atom bomb is said to have a strength of 20,000 tons. Tests of megaton devices have been reported. The energy of explosion is released in a very short time, of the order of a microsecond.

The radiation effect of a nuclear explosion is extremely severe at distances up to a few kilometers. Table 24.2 shows the distances at which neutron dose of 500 rem is received for different yields.

Table 24.2.   Distance–Yield Relation for Nuclear Explosion.

Yield (tons)	Radius (meters)
1	120
100	450
10,000	1050
1,000,000	2000

Special designs of devices have been mentioned in the literature. Included are "radiological weapons" intended to disperse hazardous radioactive materials such as Co-60 and Cs-137. More recently the "neutron bomb" has been invented, which is a small thermonuclear warhead for missiles. Exploded at heights of about 2 km above the earth, it has little blast effect but provides lethal neutron doses.

24.2 NONPROLIFERATION

We now take up the subject of proliferation of nuclear weapons and the search for means to prevent it. To reduce the spread of nuclear materials has recently become more important as the result of increases in political instability and acts of violence throughout the world. The topic is one involving many paradoxes, as we see in the following.

The international aspect of nuclear weapons first appeared in World War II when it was believed by the Allies that Germany was well on its way to producing an atomic bomb. The use of two weapons by the United States to destroy the cities of Hiroshima and Nagasaki alerted the world to the terrible consequences of nuclear warfare. Concerned about the effects of radioactive fallout from the atmosphere resulting from nuclear weapon testing, a Limited Test Ban Treaty among several nations was signed in 1963. This treaty permitted only underground tests.

In 1968 an international treaty was developed at Geneva with the title Non-Proliferation of Nuclear Weapons (NPT). The treaty is somewhat controversial in that it distinguishes states (nations) that have nuclear weapons (NWS) and those that do not (NNWS). The main articles of the treaty require that each of the latter would agree (a) to refrain from acquiring nuclear weapons or from producing them, and (b) to accept safeguards set by the International Atomic Energy Agency, based in Vienna. The treaty involves an intimate relationship between technology and politics on a global scale and a degree of cooperation hitherto not realized. There are certain ambiguities in the treaty. No mention is made of military uses of nuclear processes as in submarine propulsion, nor of the use of nuclear explosives for engineering projects. Penalties to be imposed for noncompliance are not specified, and finally the authority of the IAEA is not clear. The treaty has been signed by 102 nations, with notable exceptions France and China as NWS. India was a signatory as NNWS but proceeded to develop and test a nuclear weapon.

The nuclear weapons states (NWS) such as the U.S. and the U.S.S.R. can withhold information and facilities from the nonnuclear weapons states (NNWS) and thus slow or deter proliferation. To do so, however, implies a lack of trust of the potential recipient. The NNWS can easily cite examples to show how unreliable the NWS are.

To prevent proliferation we can visualize a great variety of technical modifications of the way nuclear materials are handled, but it is certain that a country that is determined to have a weapon can do so. We also can visualize the establishment of many political institutions such as treaties, agreements, central facilities, and inspection systems, but each of these is subject to circumvention or abrogation. It must be concluded that non-proliferation measures can merely reduce the chance of incident.

We now turn to the matter of employment of nuclear materials by organizations with revolutionary or criminal intent. One can define a spectrum of such, starting with a large well-organized political unit that seeks to overthrow the existing system. To use a weapon for destruction might alienate people from their cause, but a threat to do so might bring about some of the changes they demand. Others include terrorist groups, criminals, and psychopaths who may have little to lose and thus are more apt to use a weapon. Fortunately, such organizations tend to have less financial and technical resources.

Notwithstanding difficulties in preventing proliferation, it is widely held that strong efforts should be taken to reduce the risk of nuclear explosions. We thus consider what means are available in Table 24.3, a schematic outline.

Table 24.3. Nonproliferation Measures.

```
        Technical                                          Institutional
          /    \                                             /      \
           \                  Safeguards                   /
            \                /          \                 /
Alternative cycles      Equipment      Procedures      Treaties
       |                    |              |               |
e.g., coprocessing     e.g., detectors,  e.g., material   e.g., nonproliferation
of U and Pu            barriers          accounting,      of nuclear weapons
                                         guarding
```

24.3 SAFEGUARDS

Protection against diversion of nuclear materials involves many analogs to protection against the crimes of embezzlement, robbery, and hijacking. Consider first the extraction of small amounts of fissile material such as enriched uranium or plutonium by a subverted employee in a nuclear facility. The maintenance of accurate records is a preventive measure. One identifies a material balance in selected process steps, e.g., a spent-fuel dissolver tank or a storage area. To an initial inventory the input is added and the output subtracted. The difference between this result and the final inventory is the material unaccounted for (MUF). Any significant value of MUF prompts an investigation. Ideally, the system of accountability would keep track of all materials at all times, but such detail is probably impossible. Inspection of the consistency of records and reports is coupled with independent measurements on materials present.

The restriction on the number of persons who have access to the material and careful selection for good character and reliability are common practices. Similarly, a limitation on the number of people who have access to the records is desirable. It is easy to see how falsification of records can cover up a diversion of plutonium. A discrepancy of only 10 kg would allow for material for one weapon to be diverted. Various personnel identification techniques are available such as picture badges, access passwords, signatures, fingerprints, and voiceprints.

The usual devices of ample lighting of areas, the use of a guard force, burglar alarms, TV monitoring, and barriers to access provide protection against intruders. More exotic schemes to delay, immobilize, or repel attackers have been considered, including dispersal of certain gases that reduce efficiency or of smoke to reduce visibility, and the use of disorienting lights or unbearable sound levels.

Illegal motion of nuclear materials can be revealed by the detection of characteristic radiation, in rough analogy to metal detection at airports. A gamma-ray emitter is easy to find, of course. The presence of fissile materials can be detected by observing delayed neutrons resulting from brief neutron irradiation.

In the transportation of strategic nuclear materials, armored cars or trucks are used, along with escorts or convoys. Automatic disabling of vehicles in the event of hijacking is a possibility.

24.4 POTENTIAL CIVIL ENGINEERING APPLICATIONS

Soon after the development of nuclear explosives it was realized that they had potential for peaceful uses and many studies were made in the program called "Plowshare."† Ideas included large-scale excavations of earth or rock, and stimulation of natural gas and heat energy releases. Provisions of the Limited Test Ban Treaty of 1963 include a prohibition of transfer of radioactivity from explosion across country borders, limiting the extensive use of the concept. It remains a potential source of benefit as needs for water, energy, and transportation routes increase.

To appreciate these applications, consider the technology of underground explosives. In a typical test, a hole is drilled several thousand feet deep, the thermonuclear device is lowered to the bottom of the shaft, and the fission–fusion reaction is set off. The amount of energy release can be predetermined by the construction of the device. Suppose, for example, the total energy release is equivalent to that from 100 kilotons of TNT. Of this, 1% might be fission energy, 99% fusion energy. Detonation produces a shock wave, consisting of material moving outward at uniform speed into the surroundings. Since the shock is composed of an ionic plasma at extremely high temperature, it vaporizes, melts, crushes, dis-

†*Holy Bible*, Isaiah 2-4 and Micah 4-3, "And he shall judge among the nations, and shall rebuke many people: and they shall beat their swords into plowshares, and their spears into pruning hooks: nation shall not lift up sword against nation, neither shall they learn war any more."

places, or cracks the rock as the energy is dissipated. A large cavity in the previously solid rock is produced. For example, a nuclear explosive equivalent to 300 kilotons of TNT buried to a depth of 1200 ft creates a spherical cavity of about 170 ft in diameter, largely filled with broken rock and gases at temperatures of several thousand degrees Celsius. Figure 24.3 shows the result of such an explosion.

Fig. 24.3. Underground cavity caused by fusion explosion. (Courtesy of Lawrence Research Laboratory, Livermore, and the United States Atomic Energy Commission.)

The character of the cavity depends on the placement of the charge. It has been found that the volume of cavity produced is directly proportional to the energy release and varies inversely (approximately) with the weight of the column of rock above the point of detonation, i.e., the deeper the shot, the smaller the cavity. Figure 24.4 shows schematically the effect of a deep underground explosion. Extending vertically upward from the shot point is a column of cracked or broken rock. This space, called a "chimney," is often several times the diameter of the cavity.

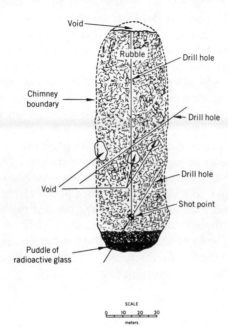

Fig. 24.4. Effect of underground nuclear explosion.

24.5 NATURAL GAS STIMULATION

One proposed use of underground nuclear explosives is for the stimulation of natural gas production. The mechanism for improvement is readily understood. In a conventional well, the gas flows from the slightly permeable rock of an underground reservoir to the well bore, which is about 6 in. in diameter. When a nuclear explosion takes place deep in the ground, the region of broken rock, many hundreds of feet in diameter, now becomes the new effective well bore. The collection area is thus multiplied by a large factor.

The two initial experiments aimed at stimulation of gas production were Project Gasbuggy, 29 kilotons at 4240 ft in New Mexico, 1967, and Project Rulison, 42 kilotons at 8426 ft in Colorado, 1969. Continued measurements of gas pressure and flow showed an improvement in production.

Accompanying the explosions were radioactive products and the long-term usefulness of the process depends on control of the amount of activity that gets into the gas. It appears that fission products such as iodine are trapped on rock surfaces, but that there are significant amounts of Kr-85 and tritium. The tritium is the isotope of greatest concern. However, it will appear only in the initial gas, which can be dispersed with dilution, or which can be burned in a controlled way to collect the tritium. Alternatively, an explosive based on fusion only could be used.

As the reserves of natural gas decline and the price increases, nuclear stimulation methods may become increasingly attractive. The gas could be pumped out or stored in the cavity for future use as needed. It has been estimated that a stimulated well would produce about 15 times that of a conventional well. Under these conditions, the process might be economically feasible, even though each 100-kiloton device costs around half a million dollars. It has been predicted that the United States reserves of around 2×10^{11} ft^3 could be doubled by successful application of the nuclear method to presently nonproductive gas basins.

24.6   NUCLEAR EXCAVATION

There are many potential uses for nuclear explosives in civil engineering works. Over the years since 1957 when the Plowshare program began, engineering studies have been made on the feasibility and economic benefit of projects such as a new canal between the Atlantic and Pacific oceans in Panama, Columbia, or Nicaragua, the excavation of harbors in Alaska and Australia, cuts in California mountains to accommodate railroad and highway traffic, a canal to connect two rivers in the South, and the production of large quantities of rock for a dam in Idaho. Several of these were found to be technically possible, but for various reasons have not been carried out.

For nuclear excavation projects, charges are buried near the surface of the earth, and the explosions produce craters whose size and shape depend on the energy and placement of the explosive and on the geological formation. A great deal of knowledge is available on these effects. More information is needed, however, on the possibility and extent of damage by earthquake induced by a series of nuclear explosions.

24.7   EXTRACTION OF OIL, MINERALS, AND HEAT

Consideration has been given to the possibility of using nuclear explosives for a variety of other-purposes.

(a) Large amounts of oil are present in oil shale, which contains solid hydrocarbons. It has been proposed to break up the shale rock by a nuclear explosion. Subsequently, air would be supplied to the cavity to maintain a fire that heats and decomposes the hydrocarbons to release liquid oil.

(b) The mining of ores such as copper is not economical by conventional means when the deposits are too deep. An explosion would create a cavity filled with ore-bearing rock, and acid would be pumped in to extract the mineral.

(c) The existence in the earth of large amounts of stored heat energy has long been recognized. In some locations, steam is readily available for generation of electricity. Nuclear explosives might be used to make such geothermal power available in many places. Where there is a high enough temperature of the rock formation, an explosion would break up the rock, and water would be piped in to be converted into steam, as sketched in Fig. 24.5.

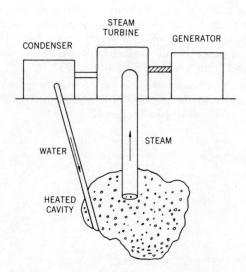

Fig. 24.5.   Explosion-stimulated geothermal power.

In all such potential applications, the effect of radioactivity must be carefully considered. The slow decay of composite fission products has been described earlier (Chapter 20). Among the fission products of primary concern are those of intermediate half-life such as Sr-90 and Cs-137, while the fusion reaction yields tritium and C-14.

24.8 SUMMARY

Nuclear explosions involve fission, using "gun" and "implosion" methods. Others use fusion of hydrogen isotopes, with very large energy and radiation yields. Concerns exist about the proliferation of nuclear weapons among countries of the world and a variety of technical and institutional measures are employed. Nuclear explosives have the potential to perform useful works such as excavation and resource stimulation.

24.9 PROBLEM

24.1. A proposal is advanced to pipe the heat of fusion explosions from a cavity deep in the earth up to the surface. If no energy were lost, how often would a 100-kiloton device have to be set off in order to obtain 3000 MW of thermal power? What would be the cost of the explosives per year? How does this "fuel" cost compare with that from fission (see Section 13.3)?

24.2. Imposion of a sphere reduces its radius to 0.8 of its original value. What is the ratio of final and original densities? Of final and original neutron mean free paths? Of geometric bucklings B^2 (see Section 12.3)?

25

Alternative Nuclear Power Systems

There was only one choice of reactor concept available to Fermi and his co-workers — a natural uranium, graphite-moderated reactor. In the period 1950–1960, a variety of types were investigated for possible commercial application, including the homogenous aqueous reactor, the organic moderator-coolant reactor, and the bismuth-cooled fast reactor. The light-water reactor suddenly became the main type adopted in the U.S. and abroad and the others sank into obscurity. More recently, there has been a revival of interest in alternative concepts besides the LWR, for reasons to be discussed.

25.1 THE ROLE OF OPTIONS

In considering the choices related to nuclear energy use, we start with the organization that is responsible for supplying electric power to a region. It may be a public or private power company regulated by a political unit such as a state or may be a branch of the government of a nation. A sequence of decisions must be made by the particular institution:

(a) Whether or not to expand the electric power capacity, dependent on anticipated or desired industrial growth and domestic usage.

(b) Whether or not to consider the installation of a nuclear plant in preference to a fossil fuel plant or a dam and hydroelectric station.

(c) Whether or not adequate long-term nuclear resources — natural

261

uranium, enrichment capacity, enriched uranium, or thorium — are available within the area or by importation of the raw material, enriched material, or the technology to process.

(d) Whether or not it is desirable to purchase a completed nuclear plant or to design and construct much of the facility with available manpower.

(e) What type of reactor to install in light of finances, technical knowledge, natural resources, concern for safety, and the climate related to nuclear weapons proliferation.

Since the largest number of reactors in the U.S. and in the world are of the light water type — PWR or BWR — this type would be likely to be examined first.

The light water reactor has a number of demonstrated virtues — the moderator is inexpensive, the fuel form provides excellent retention of fission products in normal operation, and the reactor operates smoothly and safely. The LWR has a few disadvantages, however. It requires slightly enriched fuel, thus reducing the efficiency of use of natural uranium resources, since a considerable amount of U-235 goes into the tails stream of the isotope separator. The enrichment facilities are costly to build and need significant amounts of energy to operate. The LWR operates primarily on U-235 that is initially of concentration about 3%. At the end of the fuel exposure, the concentration is less than 1%. The conversion ratio is relatively small, around 0.6, and only 5% of the uranium is consumed. Then, since the product of the isotope separator is only one-sixth of the feed, the overall utilization of natural U for power is about 1%. However, this implies that close to half the U-235 in the natural U feed is burned for power production. The rest is left as depleted tails or spent fuel. Thus, the rate of supply of mined uranium to fuel the LWR's is excessive. On the basis of known fuel reserves and expected growth in numbers of light water reactors, it is predicted that low-cost uranium (less than \$30/lb of U_3O_8) will be exhausted shortly after the end of the twentieth century.

In spite of the low conversion ratio, there is sufficient production of plutonium in the LWR fuel assemblies to make it necessary to provide an extensive safeguard system in the fuel cycle, to assure that the fissile and hazardous plutonium is not diverted for illegal use. In the interests of speed of licensing and construction of reactors, it is desirable to standardize on a single type. The cost of fabrication and operation is obviously lower if systems are mass-produced.

25.2   ALTERNATIVE FUEL CYCLES

When one considers all of the possible combinations of nuclear fissile and fertile fuels, the moderators and coolants, and the arrangements of the different materials, the number of different fuel cycles is enormous. In the study labeled Nonproliferation Alternative Systems Assessment Program (NASAP) performed by the U.S. Department of Energy, some sixty-seven different systems are identified, and variants of these would lead into thousands of candidates. We shall merely outline the scope of the choices, emphasizing the key distinctions, based on our summary of reactor concepts in Chapter 13. The problems of fuel resources and proliferation have led to the exploration of alternative systems that might be preferable to conventional LWR converters from the standpoint of natural resource conservation, economy, safety of the public, and protection against proliferation.

Consider first some variations involving the basic LWR. The simplest mode of operation, Fig. 25.1, is to supply slightly enriched U to the fuel fabrication plant and to store the spent fuel assemblies in a "once-through" mode. Next, Fig. 25.2, is to separate out the uranium in spent fuel by partial reprocessing, leaving plutonium mixed with fission products, and to recycle the U to the isotope separator. Then, Fig. 25.3, one separates all three components of spent fuel — U, Pu, and fission products — and recycles both of the fuels, the U to the separator and the Pu to the fuel fabrication facility to form the "mixed oxide" fuel assemblies. The effect of recycling is to reduce the demand on natural uranium

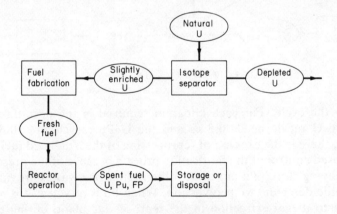

Fig. 25.1. Once-through cycle.

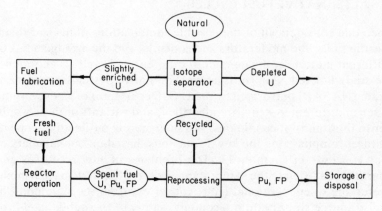

Fig. 25.2. Recycled uranium.

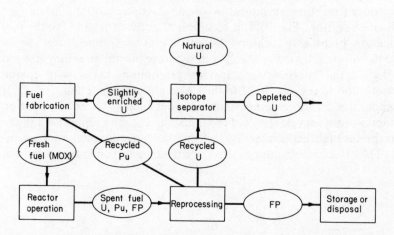

Fig. 25.3. Recycled U and Pu.

supply to the cycle. For each kilogram required in the once-through mode, the U recyle needs 0.8 kg and the U–Pu recycle only 0.6 kg. However, there is the expense of reprocessing of the irradiated fuel and the increased control of the fabrication process because of the presence of Pu. Leaving stored Pu mixed with the radioactive wastes provides a considerable deterrent to its diversion. When Pu is recycled, it is more accessible to illegal extraction in the reprocessing plant, during transportation, and in the refabrication process. Several schemes designed to prevent this have been suggested. One is to recycle plutonium mixed with some or all of the fission products in a "spiked" mode. Another is to recycle plutonium coprocessed with uranium, i.e., still mixed with an appropriate amount of uranium, which serves as a diluent for the use of Pu in a weapon. Even greater deterrent value is the "CIVEX" process in which U, Pu, and fission products are recycled together.

There are ways to minimize the penalty incurred with once-through operation, which leaves unused U and Pu. One is to increase the neutron exposure for LWR fuel assemblies by leaving them in the reactor longer. This requires slightly thicker cladding to assure integrity of the fuel tubes and a higher initial enrichment to compensate both for the increased structure and the greater depletion. Ideally, less Pu would be left in the assemblies after irradiation.

Alternatively, one can turn to the heavy water moderated reactors of the CANDU type which operate essentially on natural uranium fuel which burns down to a disposable U-235 content of about 0.3 percent. The plutonium concentration is not very large at the end of life and disposal does not pose a serious penalty. The advantage of elimination of isotope separation is partly balanced by the high cost of the moderating heavy water.

The timely introduction of a breeder reactor is a partial solution of the resource problem. The successful commercialization of a liquid metal cooled fast breeder reactor that uses and produces plutonium would cut down the demand for natural U and extend the period of use of low-cost uranium to the order of a thousand years. Of course, the plutonium inventory would be very large and the chances of weapons proliferation would be increased. In the event of a nuclear accident, the hazard to the public would also be greater. On the other hand, it is argued that the installation of plutonium in a reactor core, which is inherently highly radioactive, is preferable to accumulating and storing plutonium in some possibly accessible manner.

25.3 THORIUM REACTORS

The use of thorium opens up an entirely new set of possibilities. It is believed that there exists in nature at least as much thorium as uranium, and use of thorium as fertile material would stretch the supply of low cost fuel. The fact that Th-232 does not have an accompanying fissile isotope means that separated U-235 or bred Pu-239 or U-233 must be supplied. The classic system involving thorium is the high-temperature gas-cooled reactor (HTGR) using graphite moderator and helium coolant. There is a general impression that the HTGR is as safe or safer than the LWR for a variety of reasons related to heat capacity and coolant characteristics. Because of the need for highly enriched uranium, the HTGR is less attractive from the proliferation standpoint, but there is a good possibility that the enrichment could be reduced to a level that is less favorable for weapons fabrication.

Another promising thorium system is the light water breeder reactor (LWBR), a test version of which is installed in the Shippingport, Pennsylvania, plant. The core consists of twelve hexagonal fuel modules, each with two regions. The center or seed region contains 6 weight percent $^{233}UO_2$ in ThO_2, and is movable to provide control during operation without loss of neutrons to extraneous absorbers. The outer or blanket region contains only 3 weight percent $^{233}UO_{22}$ in ThO_2, and serves as a breeding region. The core, moderated and cooled with water, is designed to be substituted for a conventional PWR core.

The cycles using thorium and U-233 involve a radiation problem that increases costs of fuel fabrication but is advantageous for proliferation protection. By several routes as sketched in Fig. 25.4(a) U-232 is formed

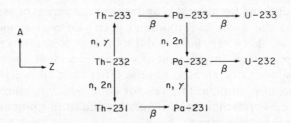

Fig. 25.4(a). Formation of U-232.

that has a half-life of 72 years. It decays to Th-228, $t_H = 1.9$ years, which yields after many steps, see Fig. 25.4(b), the isotope Tl-208, a 2.6-MeV gamma emitter. Shielded remote-controlled fabrication facilities are required if there is appreciable U-232 contamination of the U-233. The high radiation level would deter would-be thieves effectively, of course.

If a great deal of U-233 were produced by converters, advanced converters, or breeders, it is possible that the isotope could be mixed with

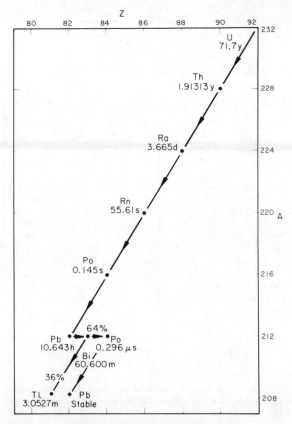

Fig. 25.4(b). Decay of U-232.

natural or depleted uranium to form a "denatured" fuel comparable in value to slightly enriched uranium. To effect the diversion of U-233, the fissile isotope would have to be separated isotopically rather than chemically. Other concepts that have been advanced include the burning of plutonium in a thorium reactor to produce U-233, in effect exchanging a less desirable fissile isotope for a more desirable one. Even more generally, one can visualize the coupling of reactor concepts to utilize all of the uranium and thorium to produce energy as in Fig. 25.5.

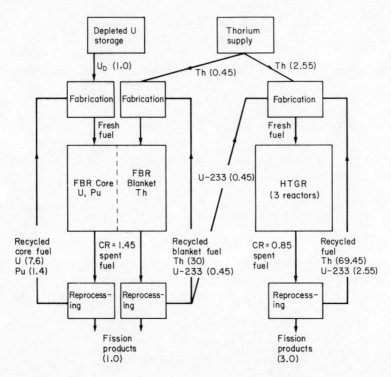

Fig. 25.5. Symbiosis of fast breeder reactor and high temperature gas-cooled reactor (numbers are kg per 1000 MWd). Adapted from Melese-d'Hospital and Simnad in *Energy* September 1977.

25.4 NUCLEAR CENTERS

There are advantages to locating a variety of components at a few sites called energy centers. These would have several reactors and their supporting facilities for fuel enrichment, fabrication, reprocessing, and even waste disposal. As many as ten reactors might be located in one area. The first advantage is the economical use of land, since the surrounding exclusion area is common to several facilities. A related benefit is the reduction in risk to the population of radioactive contamination. Second is the improved ability to provide protection against intrusion, sabotage, and theft of material since lines of transport of sensitive material are very short. The possibility of diversion of nuclear materials for illegal use would be minimized. Third, the staff responsible for the site could be more highly skilled and trained than those of many dispersed sites. Finally, there is the possibility of use of a resident international inspection team to detect activity related to proliferation of weapons.

A variant on the energy center is the regional fuel cycle center which contains the parts related to strategic fissile materials and separated radioactive wastes. Included would be spent fuel storage, reprocessing, plutonium storage, mixed U–Pu oxide fuel fabrication, and waste disposal. Enrichment facilities for production of typical 3% U-235 fuel and the plant for fabrication of this slightly enriched fuel into assemblies could be located outside the center. Reactors located at various places over the country would ship the highly radioactive spent fuel to the center for processing with little chance of its being diverted.

As for all concepts, some disadvantages can be identified. The demand for water for cooling in energy centers may be excessive and the local environmental impact might be intolerable. The consequences of any disruption of a center's service would be more severe because of the high concentration of electrical production.

25.5 OFFSHORE AND UNDERGROUND NUCLEAR PLANTS

In order to achieve improved safety and acceptability for nuclear power plants, several novel siting concepts have been proposed, including the use of islands, the seabed or beneath it, and satellites. Two more realistic ideas are "offshore" and "underground."

Offshore Power Plants

This concept involves mounting the nuclear electric generating station on a barge platform that floats in the ocean several miles from the coast. A surrounding breakwater provides protection from waves and collisions with ships. Advantages of such an arrangement are the availability of new and inexpensive sites that are close to load centers, insulation from earthquake shock, abundance of cooling water for removal of waste heat, and remoteness from population centers in event of release of radioactivity. The system would be prepared at a shipbuilding facility and towed to its eventual location. Disadvantages of the idea include the need for long flexible submerged power cables, the potential for fires following collision with an oil tanker, and unwanted aesthetic intrusion. Installation of such a plant off the coast of New Jersey is planned by Westinghouse Electric Corporation and Tenneco, Inc.

Underground Nuclear Plants

Consideration has been given to locating nuclear stations partly or completely below the earth's surface. Various schemes include back-filling of soil around a surface plant, covering over a plant, covering over a plant placed in a cut, and locating the facility deep in a rock cavern. The main advantages cited are the additional protection of the public in case of reactor meltdown and radioactivity release, improved resistance to seismic activity, reduced possibility of groundwater contamination, and improved aesthetics. Plants could be built closer to the place where they are needed. Disadvantages are the increased capital cost (~25%) related to construction, the possibility of accidental flooding with water, and difficulties with ventilation. Although several reactors have been built underground in Europe, there is no experience with large modern power reactors. Several engineering studies have been made as described in the references (see Appendix).

25.6 ACCELERATOR–BREEDER

Studies have been made on the use of a charged-particle accelerator to produce neutrons that are absorbed by fertile material in a reactor. Early work, 1949–1954, was at Livermore in the MTA project. When high-energy (500 MeV) protons or deutrons bombard a material such as uranium the processes of spallation, neutron evaporation, and fast fission give rise to large numbers (up to 30) neutrons. Surrounding the target of a linear accelerator (recall Section 9.3) nuclear fuel can be loaded to make use of the neutrons. Several modes of operation of the reactor have been suggested (a) as a subcritical reactor driven by the external neutron source, producing power without the possibility of excursions, (b) as a breeder, in which the extra neutrons permit a steady-state cycle with natural uranium feed only, or (c) as a regenerator of spent fuel, where new plutonium is created, effectively to re-enrich the fuel. In the latter mode one could extend fuel usefulness without repro-cessing.

Some of the goals that would have to be met in an accelerator–breeder are the achievement of steady reliable accelerator operation, the safe dissipation of target heat (possibly by the use of a liquid metal target) and the redesign of fuel assemblies to assure freedom from cladding failure under extended exposure. Preliminary analysis indicates that the energy released in the target is several times the beam energy and that much of the target energy can be recovered. Thus a relatively small amount of power from the outside is required.

25.7 SUMMARY

Motives for considering alternatives to the light water reactor are (a) improvement in uranium utilization and (b) the discouragement of nuclear weapon proliferation. Possibilities are other fuel cycles, thorium reactors, and new siting arrangements. The accelerator–breeder involves production of neutrons by high-speed charge particle bombardment.

25.8 PROBLEMS

25.1. Sketch the following fuel cycles as adaptations of Figs. 25.1–25.3:
  (a) spiked Pu recycle,
  (b) coprocessed U-Pu recycle,
  (c) coupled LWR and HTGR (with fissile Pu, fertile Th).

25.2. The number of atoms of a radioisotope that is produced by decay of a precursor is given by

$$N_2 = N_1^0 \, \lambda_1 \frac{e^{-\lambda_1 t} - e^{-\lambda_2 t}}{\lambda_2 - \lambda_1}$$

Apply this relation to find the activity of thorium-228, $t_H = 1.913$ yr, as the daughter of uranium-232, $t_H = 71.7$ yr, at one year from the production of 1 gram of U-232.

25.3. Deutrons are given an energy of 500 MeV in a breeder accelerator. If the beam current of D^+ ions is 0.25 amperes, how much is the target power? Noting the charge per ion of 1.60×10^{-19} coulombs, and the yield of 24 neutrons per deuteron, find the neutron production rate in the target.

26

Thermal Effects and the Environment

The generation of electrical power by consumption of any fuel is accompanied by the release of large amounts of waste heat. As the demand for power grows, the potential effect on the environment is a matter of increasing concern. In this chapter we review the physical basis for the waste energy release, describe the available mechanisms for dealing with thermal discharges, and consider ways of turning the energy to beneficial purpose.

26.1 THERMAL EFFICIENCY

Let us examine the origin of waste heat from a plant that generates electrical power. A great deal of steam is passed through a turbine to provide the mechanical energy that drives the electrical generator, as discussed in Chapter 14. The steam leaves at a low temperature, e.g., 49°C (120°F). However, because of the latent heat of vaporization it has a large heat content, which is given up when the steam is condensed and returned as water to the steam generator. A large flow of water at a temperature near that of the surroundings must thus flow through the condenser to remove this thermal energy, which constitutes a large part of the waste heat.

For any energy conversion process the thermal efficiency e, defined as the ratio of work done to thermal energy supplied, is unfortunately limited by the temperatures at which the system operates. In accord with the

272

second law of thermodynamics, in an ideal cycle, the highest efficiency is

$$e = 1 - \frac{T_1}{T_2},$$

where T_1 and T_2 are the lowest and highest absolute temperatures.[†]

To illustrate, suppose that a steam generator produces steam at 300°C and the cooling water comes from a reservoir at 20°C. The maximum efficiency for this part of the system is then

$$e = 1 - \frac{293}{573} = 0.49.$$

The overall efficiency of the plant is lower still because of effects in piping, pumps, and turbine, reducing the efficiency for a nuclear plant to around 0.33. This means that twice as much energy is wasted as is converted to useful electrical energy. Even if we were able to use water at the freezing point 0°C, little improvement in efficiency would be achieved. A large gain would result, however, from increasing the steam temperature, say to 500°C. Such higher steam temperatures are possible in a fossil fuel plant, giving overall efficiencies of around 0.40. The difference between 33% and 40% appears small, but for a given electrical generating capacity, say 1000 MWe, the waste heat for a typical light water nuclear plant is some 35% higher than for a modern fossil fuel plant. There are good indications that the gas-cooled thermal reactor and the liquid metal fast breeder reactor will achieve efficiencies comparable to fossil fuel plants.

26.2 HEAT REJECTION METHODS

The effect on the environment of discharging thousands of megawatts of waste heat from a generating station is intimately related to the manner in which the energy is dissipated. In order to understand the magnitude of the problem of handling condenser cooling water, let us calculate the flow needed, using simple energy considerations. Assume that the efficiency of a 1000-MWe plant is 0.33, so a thermal power $P = 2030$ MW must be dissipated. If this is absorbed by the condenser cooling water, specific heat $c = 4.18$ J/g-°C, to give a temperature rise of typically $\Delta T = 12$°C, the mass flow rate must be

$$M = \frac{P}{c\Delta T} = \frac{2.03 \times 10^9}{(4.18)(12)} = 4.05 \times 10^7 \text{ grams per second.}$$

[†]In the Kelvin scale, these are °C + 273 or in the Rankine scale, °F + 460.

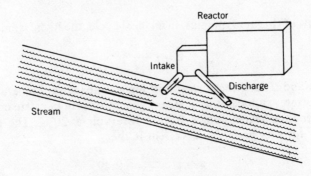

Fig. 26.1a.   Direct cooling from stream.

For reactor sites at the sea coast or next to a large inland lake, the waste heat can be readily removed by circulation and dilution. For sites near a river, the water may be extracted and returned, as sketched in Fig. 26.1a, taking advantage of natural dilution and heat transfer from the stream surface to the atmosphere to keep the temperature downstream from being excessive. From the above calculations, we see that the daily flow rate is $(4.05 \times 10^7 \text{ g/sec}) (8.64 \times 10^4 \text{ sec/day}) = 3.50 \times 10^{12}$ g/day which is also 925 million gallons per day. A flow rate of a billion gallons per day is found only in large rivers. Even complete diversion of a small stream

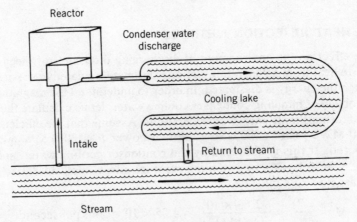

Fig. 26.1b.   Flow through cooling lake.

through the condenser would not provide the flow needed. If, instead of allowing 12°C rise, one imposes a limit on the temperature increase in a river to meet governmental regulations on water quality, say to 3°C, the stream flow would have to be at least 4 billion gallons per day.

When there is sufficient water flow in an adjacent stream but when the limit set on the increase in its temperature is too stringent, a second approach may be used. A cooling pond or lake of many acres area is constructed. As shown in Fig. 26.1b, cool water is drawn from the stream and heated water is discharged to the lake. By the time the water returns to the stream it is back to ambient temperature.

A third method is to isolate the cooling lake from the public waters, with the condenser cooling water drawn from and returned to the lake (Fig. 26.1c). The thermal energy deposited therein is released to the atmosphere by convection involving air currents over the surface, by evaporation, and by radiation. Makeup water from a public source is required only during periods in which rainfall does not replenish the water evaporated. We can find the upper limit on the water makeup if all heat

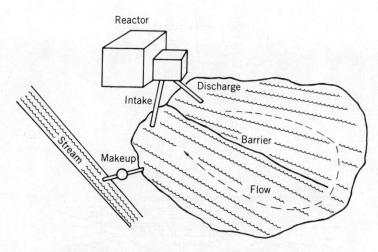

Fig. 26.1c.   Separate lake, makeup from stream.

loss were by evaporation. For a mass M of water with specific heat c, to reduce the temperature by ΔT requires energy $Mc\Delta T$. This comes from evaporation of a mass m with heat of vaporization L, energy mL. Equating.

$$\frac{m}{M} = \frac{c\Delta T}{L}$$

using $c = 4.18$ J/g-°C, $\Delta T = 9$°C, and $L = 2.26 \times 10^3$ J/g, we find $m/M = 0.0167$, which says that nearly 2 percent of the day's cooling water has to be evaporated. The higher the temperature of the discharge to a lake, the greater will be the rate of heat dissipation, and the smaller the area required. On the other hand, the effect on organisms in the water will be greater. Thus there is a competition between two desirable environmental objectives.

A fourth approach, the use of cooling towers, can be applied if land for a lake is not available or is too expensive. A cooling tower is a large heat exchanger with an air flow provided by natural convection or by blowers. In the "wet" type (Fig. 26.2a), the surface is kept saturated with moisture, and cooling is by evaporation. In the "dry" type (Fig. 26.2b), analogous to an automobile radiator, the cooling is by convection and requires greater

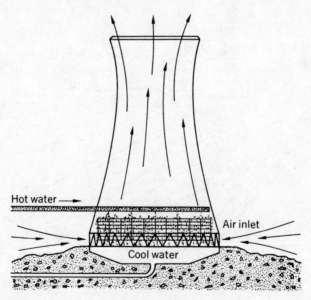

(a) "wet" (evaporative)

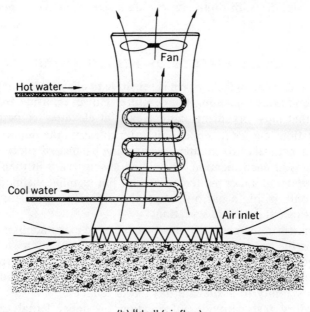

(b) "dry" (air flow)

Fig. 26.2 (a and b).   Cooling towers (From *Thermal Pollution and Aquatic Life* by John R. Clark. Copyright © March 1969 by Scientific American, Inc. All rights reserved.)

surface area and air flow. A hybrid cooling tower called "wet/dry" combines the features of the two types to minimize the effect of vapor plumes in cold weather or to conserve water during hot-weather periods. Cooling towers for a reactor plant must be very large, of the order of 100 m (300 ft) in height and of about the same diameter. Because of the large air flow required, winds of 10 m/sec (22 mph) are developed as far as 400 m from a dry tower designed for 2000 MWt removal. The wet cooling tower has a tendency to produce fog and it would not be wise to locate an airport near a reactor plant so equipped. Also, a large amount of makeup water is required to meet losses due to evaporation. It is easy to verify that the amount of water evaporated per kWh of waste heat must be about 1.6 kg or 0.42 gal. For a 1000-MWe plant this means 7.8×10^7 kg/day or 20 million gallons a day. The dry cooling tower must be larger and more expensive to build than the wet tower because it depends on convection heat removal rather than that due to evaporation. However, it has less operating and maintenance cost. It requires much less makeup water but is more sensitive to air-temperature conditions. The choice of type of tower depends on the power plant size, the environmental situation, and the climate.

26.3   BIOLOGICAL EFFECTS OF HEATED WATER

Our complex ecological system is sensitive to changes in many environmental conditions, including temperature. However, aside from stating the fact that most organisms cannot exist at elevated temperature, few generalizations can be made, and in order to assess the impact of heated water, it is necessary to examine the existing balance of species of plants and animals at each location. Individual species vary in their ability to thrive in water at abnormal temperatures. For example, salmon and trout can live only within a narrow range of temperatures while oysters and barnacles can stand wide variations.

An increase in temperature causes an increase in metabolism of animals that live in the water, doubling roughly with each 10°C increase. The need for oxygen goes up in proportion, but there is a tendency for oxygen content in water to decrease as the temperature is raised. In addition, the organism's ability to assimilate oxygen is reduced.

Laboratory tests provide information on upper lethal temperature limits. These range from 106°F for tropical fish to 77°F for salmon and trout. Some of the recommended maximum temperatures for well-being and growth are 93°F for catfish and bass, but 84°F for pike and perch; those for spawning and egg development are 75°F for bass, 55°F for salmon, and 48°F for trout.

The rates of growth of plants such as algae depend on temperature, with those causing disagreeable changes in odor or taste of water being favored by higher temperature. Also, there is an increase in the toxicity of poisons in the water, e.g., chlorine as used to clean condenser tubes, or pesticides such as DDT.

Tests on confined specimens do not reveal information on the ability of a mobile animal to avoid an adverse environment. However, if the region of excessive temperature is too large, the normal path of migration may be seriously disturbed. An increase in temperature may trigger spawning or migration at an earlier time than normal, which can affect the chances of survival.

On discharge from a power plant, heated water mixes with and is diluted by the water of the stream, lake, or ocean, and a negligible temperature rise is observed at large distances. It is recognized that there must be a mixing zone, but it is recommended that the zone be made as small as possible and not block passage for migration. It should be noted, however, that increases in temperature may be desirable, especially in colder climates. Conventional power plants have long observed the abundance of fish at the condenser discharge, presumably because of the increase in plant food supply at that point.

A legal basis has been established in the United States for the control of thermal pollution through various Acts of Congress, with responsibility for enforcement placed on the states. Recommended limits, for waters outside mixing zones, have been provided on maximum temperature and on the increase resulting from the installation of a power plant. Typical limits range from 1°C to 3°C, dependent on the season of the year and on whether the water is fresh or marine.

Much more information is needed on the complex effects of temperature on the ecology, in order to ascertain the extent of disturbance of natural conditions and the economic impact. Unfortunately, effects on the complex system are difficult to obtain without creating a significant disturbance. The imposition of an arbitrary small limit on temperature may be unnecessarily strict, unless it is assumed that our environment must not be changed in any way. Finally, political decisions must be made as to the importance of changes in the plant and animal population distribution, in comparison with power needs and the cost of preventing adverse effects.

26.4 BENEFICIAL USES OF WASTE HEAT

Much consideration has been given to the utilization of the waste heat for beneficial purposes, but relatively few large-scale applications have been initiated as yet. Some of the concepts that have been advanced are as follows:

(a) Use of the energy for home, office, and factory heating, by means of a central community system. In order to use the energy effectively, it would be necessary to plan and construct the heating and air-conditioning systems for the whole city in conjunction with power plant development. Since urban needs vary with climate and season, optimization is difficult. However, the concept has been successfully demonstrated in a small nuclear plant near Stockholm, Sweden, producing 10 MW of electricity and 80 MW of hot-water power.

(b) Enhancement of agricultural production. Warm water can be used to heat the ground and thus to stimulate growth or to increase the number of harvests of a crop per year, or to permit raising varieties that are not normally possible because of climate. Large-scale controlled production of food fish such as catfish is feasible with a heated water supply. The demand for heat varies with the season, of course, requiring alternate channels for dissipation.

(c) Desalination of salt water or brackish water. By use of the steam discharged from a turbine, desalination to produce water fit for drinking and agricultural use is feasible, with some modifications and compromises in design. The reactor system might be for the sole purpose of desalination or be a dual purpose system that optimizes for both electricity and process heat.

The concept of an agro-industrial complex built around nuclear power plants is very attractive. One can visualize the use of the electricity for manufacturing and for domestic use in the community; the discharge steam for desalination of sea water to provide water for human and animal consumption, industrial use, and for irrigation; the application of heated water for temperature control in homes and other buildings; the use of power for production of fertilizer and for pumping irrigation water; the processing of minerals extracted from the sea water; and so on.

26.5 SUMMARY

Large amounts of waste heat are discharged by electrical power plants because of inherent limits on efficiency. For typical nuclear systems, a billion gallons of water per day must be passed through the steam condenser to ensure that the temperature rise in the environment is small. The water is taken from large rivers or artificial lakes, or cooling towers are employed. Heated water can have adverse effects on plant and animal life with a great variation in sensitivity among species. Potential beneficial uses of waste thermal energy include space heating, enhancement of agricultural production, and desalination of sea water.

26.6 PROBLEMS

26.1. The thermal efficiencies of a PWR converter reactor and a fast breeder reactor are 0.33 and 0.40, respectively. What are the amounts of waste heat for a 900 MWe reactor? What percentage improvement is achieved by going to the breeder?

26.2. As sketched, water is drawn from a cooling pond and returned at a temperature 14°C higher, in order to extract 1500 MW of waste heat. The heat is

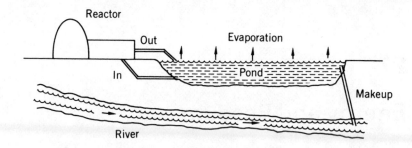

dissipated by water evaporation from the pond with an absorption of 2.26×10^3 J/g. How many kilograms per second of makeup water must be supplied from an adjacent river? What percentage is this of the circulating flow to the condenser?

26.3. As a rough rule of thumb, it takes 1–2 acres of cooling lake per megawatt of installed electrical capacity. If one conservatively uses the latter figure, what is the area for a 1000-MWe plant? Assuming 35% efficiency, how much energy in Btu is dissipated per square meter per hour from the water? Note: 1 acre = 4047 m².

2.64. Verify that about 1.6 kg of water must be evaporated to dissipate 1 kWh of energy.

27

Energy and Resources

The world has awakened recently to the realization that we are facing three related major problems — environmental pollution, the depletion of natural resources, and a population explosion. Fundamental to these is the potential shortage of energy, which is basic to meeting all of man's physical needs — clothing, shelter, transportation, convenience, and recreation, to name a few. In this chapter we shall look at the trends in energy consumption, the possible alternative sources, and the role of fission and fusion processes in man's future.

27.1 TRENDS IN ENERGY CONSUMPTION

The use of energy by a country is closely correlated with its degree of technological development and industrialization, which are in turn related to the people's standard of living. Figure 27.1 shows the past trends in energy consumption in the United States. We see that wood was the main source a hundred years ago. The growth of coal usage in the latter part of the 1800s and early 1900s is associated with rapidly increasing industrial development. There followed an enormous expansion in the consumption of natural gas and oil for heating, electrical generation, and especially transportation. The automobile had become an essential part of modern existence. Electrical energy derived from consumption of coal and oil was used for an increasing number of purposes because of its versatility, cleanliness, and ease of distribution. The widespread adoption of air conditioning for comfort, work efficiency, and health had the effect of reversing the season of peak demand for electricity from winter to summer. In the oil embargo of 1973, limits were

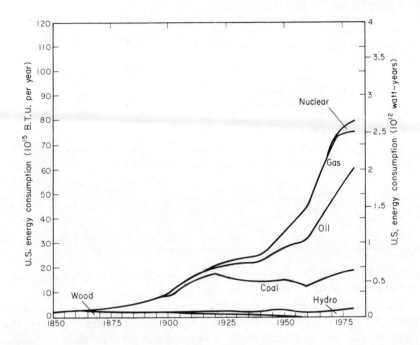

Fig. 27.1. United States energy consumption. Adapted from *Energy and Power* by Chauncey Starr. Predictions beyond 1980 are omitted because of the great uncertainties in the solution of the energy problem.

placed on shipments of oil from producing countries to consuming countries. It was generally realized for the first time that there was a world energy problem — the potential exhaustion of several natural resources.

On a world basis, the average level of industrial development and accompanying standard of living is far behind that of the United States and technologically advanced nations of Europe, and the developing countries are seeking to improve their conditions as rapidly as possible. However, the gap between developing and technologically advanced countries tends to widen, as population growth due to improved health exceeds the rate of industrialization. Dr. Chauncey Starr comments:† "If the underdeveloped parts of the world were conceivably able to reach by the year 2000 the

†In the lead article in the September, 1971 issue of *Scientific American*, which is devoted entirely to the subject of energy.

Energy usage in the United States is as listed below:

Use	Percent
Household and commercial	19
Industrial	25
Transportation	26
Electricity generation	30

The sources of energy are:

Source	Percent
Petroleum	47
Natural gas	27
Coal	19
Hydro	4
Nuclear	3

standard of living of Americans today, the worldwide level of energy consumption would be roughly 10 times the present figure."

The bulk of materials now used for energy are classified as fossil fuels — of plant and animal origin, deposited in the earth over millions of years in the form of coal, oil, and natural gas. They are classed chemically as hydrocarbons, involving the elements carbon, hydrogen, oxygen, and nitrogen. The fossil fuels are extremely useful raw materials because of the conveniently stored chemical energy, but if they are burned for fuel, the products are released and lost. Even if we disregard the pollution in the form of oxides of nitrogen, carbon, and sulfur that results from the burning, there is a staggering waste of natural resources that will never be available again. It is preferable to use the fuels for the production of more permanent and valuable products, and then to recycle them when no longer useful. Such practices are preferable to burning of wastes, which in turn is better than continued dumping of wastes.

Modern man is using the resources at such a rate that their exhaustion within a very few centuries is likely. This may seem to be a long time in comparison with a generation or even a lifetime, but in the span of man's history, it is a very brief epoch. A dramatic statement of this problem is provided by the graph in Fig. 27.2. Some interpretation of curves of this type is desirable. The use of a resource increases as exploration, extraction, and consumption increase. Eventually a point is reached at which costs of securing the resource increase toward a prohibitive level, and its use declines to nearly zero. A resource is not exhausted, it merely becomes too expensive.

27.2 ALTERNATIVE SOURCES OF ENERGY

If the world is to solve the long-range energy problem, some new practices to make better use of the available energy must be followed and new sources of energy must be discovered and developed.

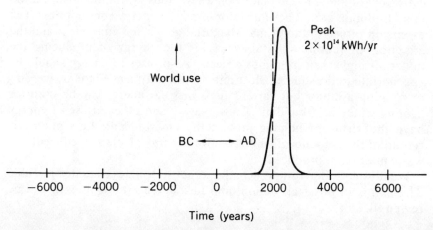

Fig. 27.2.   The epoch of fossil fuels. (Adapted from *Energy Resources: A Report to The Committee on Natural Resources*, Publication 1000-D, M. King Hubbert, National Academy of Sciences—National Research Council, Washington, D.C., 1962.)

Conservation

Restraint in the use of energy by individuals and organizations can have an important effect on the rate at which fuels are consumed. Examples are mandatory highway speed limits to reduce gasoline consumption and the maintenance of lower winter temperatures and higher summer temperatures in homes, offices, and manufacturing plants. Conscious effort to reduce energy consumption can in effect open up new resources of energy. Since such practices are personal in nature, one finds a wide range of opinion as to the efficacy of conservation measures. Their impact obviously depends on the degree with which life styles are modified. There is no doubt that conservation can be practiced in the United States and other advanced countries without serious inconvenience. There remains the problem of bringing the rest of the several billions of people of the world up to an acceptable standard of living by increased use rather than decreased use of energy.

Efficiency and Suitability of Uses

The ratio of useful work done to energy consumed is a simple definition of efficiency. Thus when efficiency is improved, the energy to do a given job is reduced. Much remains to be done by the design and application of equipment used in the home and industry. Improved insulation in all buildings would have a major effect on energy consumption. The extension of useful life of manufactured goods and equipment and the adoption of reusable containers saves the energy of producing the materials and fabrication into objects. Also, uses of energy should be appropriate to the source. Electricity is evidently needed for arc welding or running a motor but various fuels or solar energy can be used for heating water or a building. Conservation and efficient use of energy have the virtue of helping protect the environment. Less mining is required; the amounts of emissions from energy plants are reduced; less waste material is produced.

We now consider various actual alternative sources classified in Table 27.1 as fossil, physical, fission, and fusion. The status of each will be reviewed.

Table 27.1 Energy Sources.

FOSSIL	Animal and plant origin
Wood	
Coal	
Oil	
Gas	

PHYSICAL	Solar and terrestrial
Hydro	
Wind	
Tidal	
Geothermal	
Solar	

FISSION	Nuclear, heavy elements
Burner Converter Converter with Pu recycle Breeder	

FUSION	Nuclear, light elements
Deuterium–tritium Deuterium–deuterium	

Natural Gas

As a byproduct of petroleum production, natural gas has been abundant until recent years, when the rate of discovery of new wells has decreased. It is a popular source of energy because of its convenience for use and cleanliness in burning, and because of controlled low prices. The deregulation of natural gas will have the effect of stimulating new exploration and allowing ordinary economic forces to determine price. It is generally believed, however, that gas will be the first fossil fuel to be in short supply.

Some thought has been given to the use of hydrogen gas as a fuel. Evolved by the electrolysis of water, its burning is pollution-free, and it is as readily piped and as safe as natural gas. Its use does not eliminate the need for energy to effect the dissociation of water, of course.

Oil

The discovery of new reserves in Alaska and the installation of a pipeline to bring the oil out has increased the supply of oil to the United States, but is not expected to meet the long-term demands for oil. Expanded off-shore drilling for new oil will also expand reserves, with accompanying increased environmental concerns, of course. Large deposits of oil-bearing shale rock are available in the western part of the United States, but the cost of extraction of oil is high, and the increase in volume resulting from processing poses a serious disposal problem. The importation of large amounts of oil from the Middle East has many implications related to political uncertainty and international balance of payments, along with problems of adequate shipping, storage, and refining capacity.

Many observers regard the increased amount of importation of high-cost oil to the United States and Europe and other countries as a major threat to the economy of the developed world and as a serious deterrent to progress in the underdeveloped world.

Coal

U.S. reserves of coal are variously estimated to be adequate for 100–400 years, but the cost of extraction continues to increase, in part because of the introduction of badly needed mine safety measures. The effect of strip mining on the land is objectionable, and the problem of control of all gaseous emissions has yet to be solved — carbon dioxide may have effects on the atmosphere's transmission of sunlight and radioactive materials contribute to the health hazard. The development of an economic method of gasification of coal would permit the elimination of undesirable byproducts such as sulfur prior to burning. The alternative or parallel development of coal liquefaction would allow coal to be extracted from deep underground deposits. The long-range fossil fuel energy problem would still be with us, however.

Fuel cells, which convert chemical energy directly to electrical energy, have the advantage of high efficiency, reduced emission of pollutants, and ability to provide local sources of energy. Units under investigation are yet too small to make a major contribution to resource conservation.

The burning of fossil fuels to produce heat that is transformed into mechanical and then electrical energy is an inherently inefficient process, as discussed in Chapter 26. Improvements in efficiency of conventional steam plants are possible with "topping" cycles, which carry a working fluid to very high temperature. Liquid metals and gases are good candidates because of their thermal properties. Another approach to improved efficiency is the use of magnetohydrodynamic generation, in which a stream of high-temperature ionized gas replaces the moving conductor in an electrical generator, yielding direct current at high potential. As a topping cycle for conventional steam plants, an increase in efficiency from 40% to 50% may be feasible, but there are many technical and economic problems yet to be solved.

Solar Power

Energy from the sun has the potential of being a major contributor to the solution of the long-range energy problem because of its inexhaustible nature. For applications such as home space heating and water heating it is practical in many areas of the world. In the design and construction of new housing and public buildings, solar systems are appropriate. Conversion of existing dwellings remains economically marginal. Advocates of solar energy often call attention to the fact that the energy is "free" and that its use provides significant savings of fuel cost. Overlooked are some simple economic facts as can easily be illustrated. Suppose a solar device costs $3000, has a life of 15 years, and saves $250 per year in fuel bills. At first glance the net savings would be $(250)(15) = 3750$ less the initial cost, or $750. However, we should actually compare the future value of $3000 invested at, say, 6% interest i.e., $3000(1.05)^{15} = \$1790$, and the future value F of the savings A which are invested at interest rate i as they appear over a span of N years, using the relation

$$F = \frac{A\left[(1 + i)^N - 1)\right]}{i} = 250 \ \frac{\left[(1.06)^{15} - 1\right]}{0.06} = \$5819.$$

There is a net *loss* of $1371. Since solar energy cannot be stored for long periods of time, it is necessary to have a back-up electrical supply, with its own costs. Large-scale adoption of solar systems may produce certain institutional difficulties. If electric utilities are to provide the energy for such supplemental heating and cooling, they must have a large reserve capacity, which results in increased costs of the electricity. Improvements in the ability to store energy for extended periods of time would reduce the demand for new elecrical generating plants or render practical other systems. In some locations one can use pump storage, in which electricity available during periods of low demand is used to lift water into a reservoir. Subsequent flow through a hydro-electric plant helps meet high demand. The development of light, cheap, and long-life, high-capacity electric storage batteries would make practical the electric automobiles. Charging the batteries during off-peak hours would effectively serve as an energy-storage method. The conversion of electrical energy into chemical energy in the form of hydrogen gas by electrolysis of water achieves a similar purpose. Research is under way on the use of flywheels for energy storage and for smoothing out energy consumption in trains, trolleys, buses, trucks or cars. Other novel ideas include (a) off-peak heating of large insulated reservoirs of fluid which can later be circulated through a heat exchanger to generate steam and (b) the charging of a large superconducting magnet coil, which can be made to release its energy when needed.

For electrical generation using solar energy there are serious limitations. To collect and concentrate the energy by reflectors and converters of present efficiency is the major difficulty. An area of collector of about 40 km²–6 km on a side — would be needed to supply electricity to a city of a couple of hundred thousand population, with efficiencies of, say, 20%. Some problems are the wasted shadowed space, deposits on and deterioration of collector surfaces, unfavorable weather conditions, and lack of energy-storage capability. The desert is a logical place to locate solar devices because of open space, isolation, and the frequency of sunny days, but transmission costs for electrical power to urban centers would be excessive. We can hope to develop converters from heat energy to electrical energy that are more efficient than 20%, and perhaps exploit super-conductivity of metals at very low temperatures to reduce transmission costs, but it is clear that there remain many technological problems in this area.

A recently advanced concept for harnessing solar energy is intriguing. It is proposed to place collectors of the sun's rays in synchronous orbit, 22,300 miles out in space, with the energy that is absorbed by solar cells transmitted to the earth by microwaves that would easily penetrate the atmosphere. The radiation would then be collected, rectified, and transmitted through superconducting lines. The feasibility of the idea is uncertain at this time.

Other Energy Sources

Biomass is the term used for materials of plant or animal origin either in the form of a waste or as a crop grown for its energy content. They are composed of organic compounds that can be burned to produce heat, steam, and electricity, or can be converted chemically into a fuel in gaseous form, e.g., methane, or a liquid, e.g., oil or methanol (methyl alcohol). The efficiency of conversion of sunlight into useful energy is low, of the order of a percent, so the area required to supply a medium-sized (400 MWe) power plant is large — about 900 square kilometers. Use of good farming land to obtain biomass for energy competes directly with food production. Marginal lands may be suitable, but might require water and fertilizer, which require energy to obtain.

Treatment of manure and sewage can yield methane and useful residues; agricultural crop wastes can be hydrogenated to form an oil; ocean cultivation of energy-rich algae, phytoplankton, and kelp is possible. The contribution of biomass to the total energy used is small but local benefits may be high. Much research and development has yet to be done in this area.

Hydroelectric power is available when a stream can be dammed to provide a large reservoir, permitting falling water to turn a hydraulic turbine coupled to an electric generator. It is generally agreed that most of the good sites for hydropower are already in use, and that this source will not meet the total need. The pumped storage technique, in which the turbines are used to pump water up to a reservoir, helps accommodate load demands, but generates no net energy.

Geothermal power, coming from the heat in the earth's crust, is available at a few sites, notably Italy and California, where the geological formation favors the natural release of steam. The environmental aspects of this source of energy, such as saline waste-water disposal and gas emissions, are not fully known, and the number of readily accessible sites is limited. There is a possibility, as yet unexplored, that the abundant heat of the earth could be tapped at many locations by drilling deep enough.

Tidal power is a less conventional method, in which water from the sea enters and leaves a restricted channel periodically, with each stream turning a turbine. Only one full-scale plant is in existence, on the coast of France, and there are a few other promising sites where tides and terrain are suitable. For the long-range energy need, however, this approach will not be adequate.

Wind power was at one time an important source of energy on farms, and has been considered recently for use in regions where strong winds prevail. An enormous number of windmills would be required, accompanied by electric storage facilities. The aesthetic aspect of windmill power must be considered, over and beyond the technical and economic factors.

In our discussion of energy sources, a tacit assumption has been made that national growth patterns will continue indefinitely, which is impossible both practically and mathematically. It is likely that the per capita consumption, in the United States at least, will level off as most of the needs are met. There are indications also that the rate of growth of population is declining in the United States, with a possible plateau of some 300–400 million people. These trends can be significantly affected by public attitudes.

27.3 THE ROLE OF NUCLEAR ENERGY

The principal conclusion of the foregoing discussion is that every source of energy available must be developed and utilized if continued economic health of the world is to be maintained. Included is nuclear energy, in spite of potential hazard. The fraction of the total of United States electricity that is generated by nuclear reactors is around 12%.

Other sources provide percentages as follows: coal 46, oil 18, natural gas 13, and hydro 11. About seventy-five reactors with average capacity 700 MWe provide the U.S. contribution. Estimates vary greatly as to the fraction of electricity that nuclear will provide in the future, but it seems likely that the figure will be no greater than 25% by the year 2000. Whether it will be larger or smaller and indeed what the actual amount of power produced by nuclear means will be is dependent on many factors. First is the question of public acceptance of nuclear plants. If those who regard the potential hazard of accidental radioactivity release as a sufficient deterrent to continued adoption of nuclear reactors are able to convince the public of their view, the trend would clearly go to zero. If demands that additional protective equipment be developed and installed result in an excessive construction and operating cost of nuclear plants, there would be a tendency for electrical utility companies to shift back to coal-fired plants, and the relative distribution between fossil and nuclear would change greatly. If, however, concerns increase about the emissions of fossil plants and about fossil fuel shortages with sharply increased costs, or if discrepancies between energy supply and demand become serious, the growth of nuclear power might be even more rapid than anticipated.

Comparisons of costs of electrical power from the three main sources — nuclear, coal, and oil — have been made by the Department of Energy as shown in Table 27.2. It is seen that the cost of construction of a nuclear plant is high but the fuel cost is low, especially in comparison with oil. It could be argued that the relatively high capital cost of nuclear facilities makes them more vulnerable to inflation. On the other hand, the price of coal and especially oil may go up faster than other commodities.

Table 27.2. Comparison of Electrical Power Costs in the United States, in mills/kWh (capacity factor 0.65, charge 15%).

Component	Nuclear	Coal	Oil
Capital	11.9	9.2	5.9
Fuel	3.1	10.2	22.5
Operation and maintenance	2.6	1.9	1.9
Total	17.6	21.3	30.3

The relation of availability of resources and their costs is a complex subject. By means of exploration, estimates of amounts of resources can be made, but it is through actual mining that the true values are determined. However, exploration and the level of mining activity are dependent on the present and foreseen demand for the material. When the demand tends to be larger than the supply, the intensity of effort increases and vice versa. The demand is dependent on the number of nuclear reactors in operation, which is somewhat related to the degree of confidence by the industry that there will be adequate uranium at low enough cost over the life of a plant. In short, it is very difficult to predict because not only physical resources but economic and social factors are involved. Finally, predictions differ widely among experts as the result of their methods and degree of optimism. Nevertheless, in order to have a basis for further discussion, some data are reported. Figure 27.3 shows the forecasts for nuclear power for the non-communist world, as developed by the Nuclear Energy Agency and the International Atomic Energy Agency, along with predictions for the United States, provided by the Department of Energy. Figure 27.4 shows the cumulative uranium requirements (as short tons of U_3O_8) to meet the power growth expected,

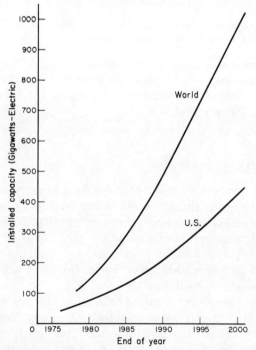

Fig.27.3. Nuclear power growth in the world and U.S. expected as of year 1978 (Source: *The Uranium Supply Outlook* by John F. Hogerton.)

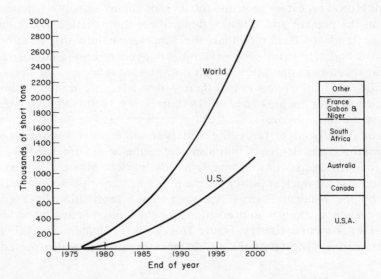

Fig. 27.4. Estimated cumulative uranium requirements and available reserves. (Source: *The Uranium Supply Outlook* by John F. Hogerton.)

both for the world and the U.S. It is assumed that neither U or Pu is recycled and that isotope separator tails have 0.25% U-235. The figure also shows in bar chart form the estimated available reserves of U_3O_8 at a price less than $30 per pound.

27.4   THE FUTURE

If man is to survive, he must plan for long-range future and take positive action to assure that there will be the necessary resources for his descendants. Several basic concepts need to be well understood by all.

1. *Terrestrial resources are finite but the need continues indefinitely.* Each resource becomes effectively exhausted at some time in the future as the cost of extracting it becomes excessive. Complete knowledge of the status of all resources is badly needed.

2. *Energy and money are required to develop new sources of energy.* Using more energy than is obtained is folly. A strong economic base is required to permit allocation of funds for research and development. If a nation becomes economically exhausted by its expenditures on materials and energy, it may never be able to recover.

3. *Renewable resources are the best for the very long term.* Wherever possible and practical solar energy should be harnessed for heating and cooling of homes and other buildings, for the generation of electricity on local or general scale, and for growing crops of plants that can be converted into alternative fuels.

4. *Transportation uses are rapidly exhausting the world's oil.* The wasteful use of automobiles, trucks, railroads, and airplanes fueled with petroleum products will be of short duration as the wells of the world dry up. Needed are strong controls and new transportation systems based on electrical energy from fuels other than fossil.

5. *Energy should be used wisely to provide basic human needs.* The uneven distribution of natural resources among nations could be alleviated by the inauguration of massive projects such as the diversion of Arctic water to the temperate and tropical zones and the construction of large-scale electrical facilities for sea water desalination and fertilizer manufacturing. The extraction of minerals from the earth by safe, continuous, automated, and inexpensive methods should be a high-priority objective.

6. *The sea is an important source of energy and materials.* There is some possibility of harnessing power from waves and temperature differences, but the main resource may be the mineral and energy content — uranium, lithium, and deuterium. Means should be found to extract uranium at a cost that would be compatible with use in breeder reactors.

7. *The time span between scientific discovery and commercial use of a major process is as long as 30 years.* This fact requires that research and development programs be authorized well in advance of need.

On the basis of these concepts we can propose a plan of action that might lead to the bright future that was visualized at the time nuclear energy was discovered.

Ample funds should be provided for the investigation of inexpensive solar systems for a variety of purposes, with financial incentives to individuals or organizations to install devices and gain experience in their use, while saving precious fuels.

Research should continue on ways to mine efficiently and to convert low-grade coal into useful chemical forms to serve as substitute for petroleum products.

A large research and development effort should be undertaken to realize practical electric automobiles and trains. Every encouragement should be provided to industry and the public to adopt the new modes of transportation. As the use grows, there would be a significant relief for importation of expensive oil for gasoline. Some of the savings of oil expenditures should be dedicated to the development of new energy sources.

In order to assure the necessary electrical energy for normal industrial growth and to power the expanding electric vehicle system, additional converter reactors should be built, using slightly enriched uranium, storing the depleted uranium for future use in breeder reactors, and using and generating some plutonium. The latter isotope could be burned as a fuel to a certain extent by recycling, with the remainder set aside for future breeder fuel.

In parallel, research and development on thorium-based reactor systems should be carried out, leading to a successful alternative source of raw material. Concepts such as the HTGR and the LWBR should be considered along with other coupled systems.

The breeder reactor should be developed with ample testing to achieve improvements to reliability, safety, and economy. It should be ready when the cost of fuel makes converter reactors uneconomical in comparison with fossil plants.

The investigation of the possibility of practical fusion reactions should be pursued vigorously, including both electromagnetic and inertial confinement methods. Consideration needs to be given to the benefits of a fusion–fission hybrid system in which plasma neutrons breed new fuel in a blanket.

Nuclear explosives should be developed that are capable of performing safely the necessary excavations to achieve abundant water supplies for all the countries of the world. Nuclear devices should also be used to create permanent disposal facilities for nuclear fission wastes deep in the earth. The use of the accumulated arsenal of nuclear weapons for these purposes would be an unprecedented act of humanity.

The solution of the energy problem should be the world's highest priority objective. If essentially unlimited energy supplies could be realized, a new era of existence for man could be imagined. The energy would be used to manufacture all of man's physical requirements, to recycle all waste materials for re-use, and to make possible new levels of technological and cultural accomplishment. If all needs are provided by abundant and inexpensive energy, harmony among nations can be visualized.

27.5 SUMMARY

A very rapid growth in energy consumption has occurred in the twentieth century, especially in the United States. The principal source—fossil fuels—is likely to be in short supply immediately and exhausted in a few hundred years, requiring that alternative sources be found. Hydroelectric, wind, tidal, geothermal, and solar power require much technological development. Conservation practices are desirable to avoid energy waste. Nuclear energy will contribute through the use of converter reactors for a number of years. Research and development leading to electric transportation backed by nuclear power is urged. The solution to the energy problem is man's highest priority.

27.6 PROBLEMS

27.1. Find the necessary collector area in square miles for solar radiation if the efficiency of conversion from thermal energy to electrical energy is 20% and the United States is provided 10^9 kW of electrical power, as is expected to be needed before 1990. Assume that the incident radiation is available 4000 hr/yr at 300 W/m². (Note: 1 mi² $= 2.59 \times 10^6$ m².) Compare the result with the area of the State of Arizona, 113,909 mi².

27.2. (a) Find the "break even" cost of a solar device, one that will provide savings that accrue to the same future value as direct investment of the initial cost using a period of 15 years, $250/yr savings, and a 6% interest rate.

(b) Find the number of years it takes for a $4000 device to pay off if the saving is $250/yr and the interest rate is 6%.

Appendix

SELECTED REFERENCES

Chapter(s)	Reference

Chapter(s)   *Reference*

General   Booklets from U.S. Department of Energy, Technical Information Center, P.O. Box 62, Oak Ridge, Tennessee 37830. (Sent free to teachers requesting on school letterhead; one or more free to students depending on their subject area; prices 1 for 45c each, 2–9 35c each, 10–30 30c each, 31 or more 20c each.)

IB-004   Computers
IB-1001   Energy Storage
IB-1002   Energy Technology
IB-701   Geothermal Energy
IB-416   Inner Space, The Structure of the Atom
IB-414   Nature's Invisible Rays
IB-1100   Oil
IB-801   Solar Energy
IB-412   Space Radiation
IB-019   Stamps Tell the Story of Nuclear Energy
IB-303   The Atomic Fingerprint
IB-405   The Elusive Neutrino
IB-018   The First Book on Information Science
IB-503   The First Reactor
IB-014   Worlds Within Worlds, Vol. 1
IB-015   Worlds Within Worlds, Vol. 2
IB-016   Worlds Within Worlds, Vol. 3

Kenneth, F. Weaver, "The promise and peril of nuclear energy," *National Geographic,* **155**, 459 (1979).

Chapter(s)	*Reference*

1-3   Raymond L. Murray and Grover C. Cobb. *Physics: Concepts and Consequences.* Prentice-Hall, Englewood Cliffs, N.J., 1970.

1-8   R. E. Lapp and L. H. Andrews. *Nuclear Radiation Physics.* Prentice-Hall, Englewood Cliffs, N.J., 1965.

5   S. F. Mughabghab and D. I. Garber. *Neutron Cross Sections Volume 1, Resonance Parameters,* BNL 325 Third Edition, Volume 1, Brookhaven National Laboratory, Upton, N.Y., June 1973.

5, 12, 13   P. J. Grant. *Elementary Reactor Physics.* Pergamon Press, Oxford, 1966.

9   M. Stanley Livingston and John P. Blewett. *Particle Accelerators.* McGraw-Hill, New York, 1962.

R. R. Wilson. "The Batavia accelerator," *Scientific American,* February 1974, pp. 72–83.

10   L. O. Love. "Electronmagnetic separation of isotopes at Oak Ridge," *Science,* October 26, 1973, pp. 343–352.

Donald R. Olander. "The gas centrifuge," *Scientific American,* August 1978, p. 37.

Richard N. Zare, "Laser separation of isotopes," *Scientific American,* February 1977, p. 86.

Simon Rippon. "A visit to Amelo," *Nuclear News,* December 1977, p. 46.

11   "Semi-conductor devices course," *IEEE Transactions on Nuclear Science,* April 1978, pp. 896–926.

William J. Price. *Nuclear Radiation Detection,* 2nd ed. McGraw-Hill, New York, 1964.

13   Hugh C. McIntyre. "Natural-uranium heavy-water reactors," *Scientific American,* October 1975, p. 17.

R. H. Renshaw and E. C. Smith. "The standard Candu 600 MW(e) nuclear plant," *Nuclear Engineering International,* June 1977, p. 45.

13   George A. Cowan. "A natural fission reactor," *Scientific American,* July 1976, p. 36.

Andrew W. Kramer. *Understanding the Nuclear Reactor.* Power Engineering Magazine, Technical Publishing Co., Barrington, Ill., 1970.

14   Summers, Claude M. "The conversion of energy," *Scientific American,* September 1971, p. 148.

15   Glenn T. Seaborg and Justin L. Bloom. "Fast breeder reactors," *Scientific American,* November 1970, p. 13.

George A. Vendryes. "Superphénix: a full-scale breeder reactor," *Scientific American,* March 1977, p. 26.

16   Terry Kammash. *Fusion Reactor Physics, Principles and Technology.* Ann Arbor Science Publishers, Inc. (1975).

George H. Miley. *Fusion Energy Conversion,* American Nuclear Society (1976).

D. Steiner. "The technological requirements for power by fusion," *Nuclear Science and Engineering* **58**, 107 (1975).

Harold P. Furth. "Progress toward a tokamak fusion reactor," *Scientific American,* August 1979, p. 50.

Chapter(s)	*Reference*
	John L. Emmett, John Nuckolls, and Lowell Wood. "Fusion power by laser implosion," *Scientific American*, June 1974, p. 24.
	Gerold Yonas. "Fusion power with particle beams," *Scientific American*, November 1978, p. 50.
	Hans A. Bethe. "The fusion hybrid," *Physics Today*, May 1979, and "The fusion hybrid," *Nuclear News*, May 1978.
17	H. D. Smyth. "Atomic energy for military purposes," *Reviews of Modern Physics*, Vol. 17, No. 4, pp. 351–471. The first unclassified account of the nuclear effort of World War II. Readily understood technical information and administrative history of the Manhattan Project. See also the book version, listed below.
	H. D. Smyth. *Atomic Energy for Military Purposes.* Princeton University Press, Princeton, N.J., 1945.
18	"The effects on populations of exposure to low levels of ionizing radiation" (the BEIR Report). National Academy of Sciences—National Research Council, Washington, D.C., November 1972.
	Daniel S. Grosch and Larry E. Hopwood. *Biological Effects of Radiations*, 2nd ed. Academic Press, New York, 1980.
18, 19	*Radiological Health Handbook.* U.S. Department of Health Education and Welfare, Rockville, Md., 1970.
20	*The Safety of Nuclear Power Reactors (Light Water Cooled) and Related Facilities.* United States Atomic Energy Commission, Washington, D.C., July 1973. Report WASH-1250.
	S. E. Rippon. "Light water reactor safety," A review of WASH-1250, *Nuclear Engineering International*, January 1974, pp. 25–30.
20, 21	R. Philip Hammond. "Nuclear power risks," *American Scientist*, March–April 1974, pp. 155–160.
	D. Bruce Turner. *Workbook of Atmospheric Dispersion Estimates*, Public Health Service. Cincinnati, Ohio, 1970.
21	Bernard L. Cohen. "The disposal of radioactive wastes from fission reactors," *Scientific American*, June 1977, p. 21.
	William P. Bebbington. "The reprocessing of nuclear fuels," *Scientific American*, December 1976, p. 30.
22, 23	Robin P. Gardner and Ralph L. Ely, Jr. *Radioisotope Measurement Applications in Engineering.* Reinhold, New York, 1967.
	Colin Renfrew. "Carbon 14 and the prehistory of Europe," *Scientific American*, October 1971, p. 63.
	W. H. Wahl and H. H. Kramer. "Neutron-activation analysis," *Scientific American*, April 1967, p. 68.
	C. L. Bennett. "Radiocarbon dating with accelerators," *American Scientist*, **67**, 450 (1979).
	J. D. Macdougall. "Fissiontrack dating," *Scientific American*, December 1976, p. 114.
	John H. Lawrence, Bernard Manowitz, and Benjamin S. Loeb. *Radioisotopes and Radiation.* McGraw-Hill, New York, 1964.

Chapter(s)   *Reference*

22–25   Glenn T. Seaborg. *Peaceful Uses of Nuclear Energy.* USAEC Division of Technical Information Extension, Oak Ridge, Tenn., 1970.

24   John S. Foster, Jr. "Nuclear weapons," *Encyclopedia Americana*, Vol. 20 (1973), p. 518.

             W. Meyer, S. K. Loyalka, W. E. Nelson, and R. W. Williams. "The homemade nuclear bomb syndrome," *Nuclear Safety*, July–August 1977, p. 427.

             Lynn E. Weaver (Ed.) *Education for peaceful Uses of Nuclear Explosives.* The University of Arizona Press, Tuscon, Ariz., 1970.

             Kevin N. Lewis. "The prompt and delayed effects of nuclear war," *Scientific American*, July 1979, p. 35.

             David J. Rose and Richard K. Lester. "Nuclear power, nuclear weapons and international stability," *Scientific American*, April 1978, p. 45.

             William Epstein. "The proliferation of nuclear weapons," *Scientific American*, April 1975, p. 18.

             Mason Willrich and Theodore B. Taylor. *Nuclear Theft: Risks and Safeguards.* Ballinger Publishing Company, Cambridge, Mass., 1974.

             Nuclear Proliferation and Safeguards. Office of Technology Assessment, Congress of the United States, Praeger Publishers, New York, 1977.

             Nuclear Proliferation Factbook. U.S. Government Printing Office, Washington, D.C., 1977.

25   George Yadigaroglu and Stephen O. Andersen. "Novel siting solutions for nuclear power plants," *Nuclear Safety* **15**, 65 (1974).

             J. H. Crowley, P. L. Doan, and D. R. McCreath. "Underground nuclear plant siting: a technical and safety assessment," *Nuclear Safety* **15**, 519 (1974).

             C. M. Van Atta, J. D. Lee, and W. Heckrotte. "The electronuclear conversion of fertile to fissile material," 1976. Report UCRL-52144.

             Proceedings of an Information Meeting on Accelerator Breeding, January 18–19, 1977, Brookhaven National Laboratory. Report CONF-770107.

26   John R. Clark. "Thermal pollution and aquatic life," *Scientific American*, March 1969, p. 18.

             Michel d'Orival. *Water Desalting and Nuclear Energy.* Verlag Karl ThiemigKG, Munich, 1967.

             Merril Eisenbud and George Gleason (Eds.) *Electric Power and Thermal Discharges.* Gordon & Breach, Science Publishers, New York, 1969.

             Riley D. Woodson. "Cooling towers," *Scientific American*, May 1971, p. 70.

             Joseph Barnea. "Geothermal power," *Scientific American*, January 1972, p. 70.

27   M. King Hubbert. "The energy resources of the earth," *Scientific American*, September 1971, p. 60.

             Resources and Man. National Academy of Sciences—National Research Council, W. H. Freeman & Co., San Francisco, 1969. Especially Chapter 8, "Energy Resources" by M. King Hubbert.

Chapter(s)   *Reference*

Chauncey Starr. "Energy and power," *Scientific American*, September 1971, p. 36.

Kenneth F. Weaver. "The search for tomorrow's power," *National Geographic*, Vol. 142, No. 5, November 1972, pp. 650–681.

Edward D. Griffith and Alan W. Clarke. "World coal production," *Scientific American*, January 1979, p. 38.

Andrew R. Flower. "World oil production," *Scientific American*, March 1978, p. 42.

H. A. Bethe. "The necessity of fission power," *Scientific American*, January 1976, p. 21.

CONVERSION FACTORS

In order to convert from numbers given in the British or other system of units to numbers in SI units, *multiply* by the factors in the following table.[†] For example, multiply the heat of fusion of water given as 80 calories per gram by 4.185 to obtain the figure of 335 J/g. Note that the conversion factors are rounded off to four significant figures.

Original system	SI	Factor
atmosphere	pascal (Pa)	1.013×10^5
barn	square meter (m²)	1.000×10^{-28}
barrel (42 gal for petroleum)	cubic meter (m³)	1.590×10^{-1}
British thermal unit, Btu	joule (J)	1.055×10^3
thermal conductivity, k (Btu/hr-ft)	W/m-°C	1.731
calorie (cal)	joule (J)	4.185
centimeter of mercury	pascal (Pa)	1.333×10^3
centipoise	pascal-second (Pa-s)	1.000×10^{-3}
curie (Ci)	disintegrations per second (d/s)	3.700×10^{10}
day	second	8.640×10^4
degree (angle)	radian	1.745×10^{-2}
degree Fahrenheit (°F)	degree Celsius (°C)	$°C = \frac{5}{9}(°F - 32)$
electron-volt (eV)	joule (J)	1.602×10^{-19}
foot (ft)	meter (m)	3.048×10^{-1}
square foot (ft²)	square meter (m²)	9.290×10^{-2}

[†]Adapted from *Standard for Metric Practice*, American Society for Testing Materials, Philadelphia, Pennsylvania (1976).

Original system	SI	Factor
cubic foot (ft³)	cubic meter (m³)	2.832×10^{-2}
cubic foot per minute (ft³/min)	cubic meter per second (m³/s)	4.719×10^{-4}
gallon (gal) U.S. liquid	cubic meter (m³)	3.785×10^{-3}
gauss	tesla (T)	1.000×10^{-4}
horsepower (hp) (550 ft-lb/sec)	watt (W)	7.457×10^{-2}
inch (in.)	meter (m)	2.540×10^{-2}
square inch (in²)	square meter (m²)	6.452×10^{-4}
cubic inch (in³)	cubic meter (m³)	1.639×10^{-5}
kilowatt hour (kWh)	joule (J)	3.600×10^{6}
kilogram-force (kgf)	newton (N)	9.807
liter (l)	cubic meter (m³)	1.000×10^{-3}
micron (μl)	meter (m)	1.000×10^{-6}
mile (mi)	meter (m)	1.609×10^{3}
miles per hour (mi/hr)	meters per second (m/sec)	4.470×10^{-1}
square mile (mi²)	square meter (m²)	2.590×10^{6}
pound (lb)	kilogram (kg)	4.536×10^{-1}
pound force per square inch (psi)	pascal† (Pa)	6.895×10^{3}
rad	gray (Gy)	1.000×10^{-2}
roentgen (r)	coulomb per kilogram (C/kg)	2.580×10^{-4}
ton (short, 2000 lb)	kilogram (kg)	9.072×10^{2}
watt-hour (Whr)	joule (J)	3.600×10^{3}
year (y)	second (s)	3.156×10^{7}

†Newton per square meter.

ATOMIC AND NUCLEAR DATA

(a) Atomic Weights (based on the atomic mass of $^{12}_{6}$C as exactly 12)

Atomic number	Name	Symbol	Atomic weight	Atomic number	Name	Symbol	Atomic weight
1	Hydrogen	H	1.00797	9	Fluorine	F	18.9984
2	Helium	He	4.0026	10	Neon	Ne	20.183
3	Lithium	Li	6.939	11	Sodium	Na	22.9898
4	Beryllium	Be	9.0122	12	Magnesium	Mg	24.312
5	Boron	B	10.811	13	Aluminum	Al	26.9815
6	Carbon	C	12.01115	14	Silicon	Si	28.086
7	Nitrogen	N	14.0067	15	Phosphorus	P	30.9738
8	Oxygen	O	15.9994	16	Sulfur	S	32.064

Atomic number	Name	Symbol	Atomic weight	Atomic number	name	Symbol	Atomic weight
17	Chlorine	Cl	35.453	61	Promethium	Pm	[145]
18	Argon	Ar	39.948	62	Samarium	Sm	150.35
19	Potassium	K	39.102	63	Europium	Eu	151.96
20	Calcium	Ca	40.08	64	Gadolinium	Gd	157.25
21	Scandium	Sc	44.956	65	Terbium	Tb	158.924
22	Titanium	Ti	47.90	66	Dysprosium	Dy	162.50
23	Vanadium	V	50.942	67	Holmium	Ho	164.930
24	Chromium	Cr	51.996	68	Erbium	Er	167.26
25	Manganese	Mn	54.9380	69	Thulium	Tm	168.934
26	Iron	Fe	55.847	70	Ytterbium	Yb	173.04
27	Cobalt	Co	58.9332	71	Lutetium	Lu	174.97
28	Nickel	Ni	58.71	72	Hafnium	Hf	178.49
29	Copper	Cu	63.54	73	Tantalum	Ta	180.948
30	Zinc	Zn	65.37	74	Tungsten	W	183.85
31	Gallium	Ga	69.72	75	Rhenium	Re	186.2
32	Germanium	Ge	72.59	76	Osmium	Os	190.2
33	Arsenic	As	74.9216	77	Iridium	Ir	192.2
34	Selenium	Se	78.96	78	Platinum	Pt	195.09
35	Bromine	Br	79.909	79	Gold	Au	196.967
36	Krypton	Kr	83.80	80	Mercury	Hg	200.59
37	Rubidium	Rb	85.47	81	Thallium	Tl	204.37
38	Strontium	Sr	87.62	82	Lead	Pb	207.19
39	Yttrium	Y	88.905	83	Bismuth	Bi	208.980
40	Zirconium	Zr	91.22	84	Polonium	Po	210
41	Niobium	Nb	92.906	85	Astatine	At	[210]
42	Molybdenum	Mo	95.94	86	Radon	Rn	222
43	Technetium	Te	[99]*	87	Francium	Fr	[223]
44	Ruthenium	Ru	101.07	88	Radium	Ra	226
45	Rhodium	Rh	102.905	89	Actinium	Ac	227
46	Palladium	Pd	106.4	90	Thorium	Th	232.038
47	Silver	Ag	107.80	91	Protactinium	Pa	231
48	Cadmium	Cd	112.40	92	Uranium	U	238.03
49	Indium	In	114.82	93	Neptunium	Np	(237)†
50	Tin	Sn	118.69	94	Plutonium	Pu	(239)
51	Antimony	Sb	121.75	95	Americium	Am	(243)
52	Tellurium	Te	127.60	96	Curium	Cm	(244)
53	Iodine	I	126.9044	97	Berkelium	Bk	(249)
54	Xenon	Xe	131.30	98	Californium	Cf	(252)
55	Cesium	Cs	132.905	98	Einsteinium	Es	(253)
56	Barium	Ba		100	Fermium	Fm	(257)
57	Lanthanum	La	138.91	101	Mendelevium	Md	[258]
58	Cerium	Ce	140.12	102	Nobelium	No	[255]
59	Praseodymium	Pr	140.907	103	Lawrencium	Lw	[263]
60	Neodymium	Nd	144.24	104			[261]
				105			[262]

*[ ] = atomic weight of most stable artificial isotope.
†( ) = atomic weight of most abundant artificial isotope.

(b) Selected Atomic Masses (rounded to six decimals)

Proton	1.007277	$^{14}_{6}C$	14.003242
Neutron	1.008665	$^{14}_{7}N$	14.003074
$^{1}_{1}H$	1.007825	$^{16}_{8}O$	15.994915
$^{2}_{1}H$	2.014102	$^{17}_{8}O$	16.999131
$^{3}_{1}H$	3.016049	$^{92}_{37}Rb$	91.91935
$^{4}_{2}He$	4.002603	$^{140}_{55}Cs$	139.91709
^{6}Li	6.015123	^{92}	235.043925
$^{7}_{3}Li$	7.016004	$^{236}_{92}U$	236.045563
$^{9}_{4}Be$	9.012182	$^{238}_{92}U$	238.050786
$^{10}_{5}B$	10.012938	$^{236}_{94}Pu$	239.052158
$^{11}_{5}B$	11.009305	$^{240}_{94}Pu$	240.053809
$^{12}_{6}C$	12.000000		

(c) Values of Fundamental Physical Constants
Selected from E. R. Cohen and B. N. Taylor, *Journal of Physics and Chemistry Reference Data*, Vol. 2, No. 4 (1973), pages 715-721.

Speed of light in vacuum, c	299, 792, 458 m/sec
Elementary charge, e	$1.6021892 \times 10^{-19}$ C
Electron-volt, eV	$1.6021892 \times 10^{-19}$ J
Planck constant, h	6.626176×10^{-34} J-sec
Avogadro constant, N_A	6.022045×10^{23} /mol
Boltzmann constant, k	1.380662×10^{-23} J/°K
Electron rest mass, m_e	9.109534×10^{-31} kg
	0.5110034 MeV
Proton rest mass, m_p	$1.6726485 \times 10^{-27}$ kg
Neutron rest mass, m_n	$1.6749543 \times 10^{-27}$ kg
Atomic mass unit, amu	931.5016 MeV
Molar volume of ideal gas, V_m	22.41383×10^{-3} m³/mol

Reference: A. H. Wapstra and K. Bos, *Atomic Data and Nuclear Data Tables*, **19**, 135 (1977).

ANSWERS TO PROBLEMS

1.1. 2400 J.

1.2. 20°F, 260°C, −459°F, 1832°F.

1.3. 2.25×10^4 J.

1.4. 512 m/sec.

1.5. 596 kWh.

1.6. 2×10^{20} /sec.

1.7. 2.2×10^{20} g.

1.8. 3.04×10^{-11} J.

1.9. 3.38×10^{-28} kg.

1.10. 3.51×10^{-8} J.

1.11. 8.67×10^{-4}.

1.13. (b) 0.140, 0.417, 0.866.

1.14. (a) 6.16×10^4 Btu/lb, (b) 1.43×10^5 J/kg, (c) 3.0 eV.

2.1. 0.0828×10^{24} /cm³.

2.2. 2200 m/sec.

2.3. 3.26×10^{15} /sec.

2.4. -1.5 eV, 4.7×10^{-10} m; 12.0 eV, 2.9×10^{15} /sec.

2.6. 8.7×10^{-13} cm, 2.4×10^{-24} cm².

2.7. 28.3 MeV.

2.8. 1783 MeV.

2.9. 1.46×10^7 kg/m³, 1.89×10^4 kg/m³, 0.99×10^{13} kg/m³.

3.1. 3.64×10^{10}/sec vs 3.7×10^{10}/sec

3.2. 1.65 μg.

3.3. 3.21×10^{14}/sec, 8.68×10^3 Ci; 1.04×10^{14} /sec, 2.86×10^3 Ci.

3.6. 4.27×10^4/sec.

3.8. 1.61×10^3 yr, radium.

4.2. $^{14}_{6}$C, $^{10}_{5}$B.

4.3. 1.19 MeV.

4.4. 4.78 MeV.

4.5. 3.95×10^{-30} kg, 3.54×10^5 m/sec, 1.3×10^{-3} MeV.

4.6. 2.05×10^7 m/sec, 1.39×10^{-12} J or 8.65 MeV.

4.7. 1.20 MeV.

5.1. 1.46 cm⁻¹, 0.68 cm.

5.2. 1.70×10^7 m/sec, 4.1×10^4/cm³.

5.3. 6×10^{13}/cm²-sec, 0.02 cm⁻¹, 1.2×10^{12}/cm³-sec.

5.4. 0.207, 0.704, 88, 0.40.

5.5. 382 barns.

5.6. 0.273×10^{13}/cm³-sec, 0.149×10^{13}/cm³-sec.

5.7. 4.43×10^{-7}, 0.443

5.8. 0.183 cm⁻¹.

6.1. 0.0233, 42.8.

6.2. 1.45×10^{21}/sec, 2.07×10^{-13} m.

6.3. 0.245 MeV.

6.4. 0.62 MeV.

6.5. ~ 0.001 cm.

6.6. 1.51 cm.

6.7. 0.289 cm.

6.8. 0.39 cm, 1.92×10^{-5} C/cm³, 2.4×10^{-4} J.

7.1. 6.53 MeV.

7.2. $^{100}_{38}$Sr.

7.3. 66.4 MeV, 99.6 MeV.

7.4. 169.0 MeV.

7.5. 2.49.

7.6. 1.0% U-235, 99% U-238.

7.7. 0.00812 g/day.

7.8. 8.10×10^6, 5.89×10^6, 5.18×10^6.

8.1. 0.02580 amu, 25.7 MeV.

8.3. 27,200 kg/day.

8.4 3.10×10^8 cm/sec, 7.28×10^{15}/cm³.

8.5. 9.3×10^5 °K.

8.6. 2.72×10^5 eV.

9.1. 0.114 volts.

9.2. 2.5×10^6/sec.

9.3. 1.31×10^{-7} sec.

9.4. $E = (qBR)^2/(2m)$

9.5. 0.183 Wb/m².

9.6. 215, 0.99999.

9.7. 750 mA, 373 MW.

10.2. 1.0030.

10.3. 0.0304, 0.0314.

10.4. 195 kg/day.

10.5. 490.

10.6. 3.944 kg/day; 0.372 SWU; 3.934 SWU, $393.40.

10.7. $n = 2[1 - \exp(-0.693\sqrt{m_H/m_L})]$ = 1.00297.

10.8. 0.281 kg/day, 0.719 kg/day

11.1. $1.21 \times 10^{21}/cm^3$, 45 times as high.

11.2. 0.0165.

11.3. 6.0×10^5.

11.4. 0.30.

11.5. 11.

12.1. 2.21.

12.2. 0.455.

12.4. $1.84 \times 10^{10}/cm^2\text{-sec}$.

12.5. 1.171, 1.033, 0.032.

12.6. 1.18, 1.84.

12.7. 2.11 weight percent.

12.8. 2.04.

13.1. 8.64×10^6, 89800 kg, 2700 kg, $70.6 million.

13.2. $11.8 million.

13.3. 2.12×10^{-5} sec, 0.037.

13.4. 133 ppm.

13.5. 156 ft^3.

13.6. 1.45 min.

13.7. 0.0331.

14.3. 3 W/cm^3-°C.

14.4. 315°C.

14.5. 30°F.

15.1. 1.7, 0.7.

15.2. 0.9856.

15.4. 2.61, 0.2.

15.5. 6300 kg, 10.6 yr.

15.7. 422,000 tonnes.

16.1. 0.1 mm, 0.65 cm.

16.4. 0.2, $500/kg, 0.003 mills/kWh.

18.1 6.25×10^4, 2.3×10^{-15}.

18.2. 200.

18.3. 1.67 mrad, 5 mrem, 0.01.

18.4. 1.8%.

19.1. 1520 mrem/yr, 360 mrem.

19.2. 56 μCi.

19.3. 1.53×10^{-9} μCi/cm^3.

19.5. 997/cm^2-sec.

19.7. 9.66×10^{-6} μCi/ml, 9.16×10^{-6} μCi/ ml, 10.26×10^{-6} μCi/ml.

20.1. 0.0157, 2.40; 7.7×10^{-4}; 63.8 sec.

20.2. 30.2 sec.

20.3. 0.0052 sec.

20.4. 40°C.

20.5. 0.90 sec.

20.6. 0.0068, 0.0046, 0.0034, 0.0021.

20.7. −0.0208.

21.1. 0–10 days I-131
10–112 days Ce-141
112 days–4.5 yr Ce-144
4.5 yr–100 yr Cs-137

21.2. 98.7%.

21.3. 1.05×10^{13} cm^3/sec, 2.22×10^{10} ft^3/min.

21.4. 30.9 days.

21.5. 1.98, 0.16, 1.21, 0.15; 0.968 tonne, 2420 MW, 80.7%.

21.6. 957 yr, 1002 yr, 801, 600 yr.

22.1 Fe-59, 6.2 days.

22.2. $^6_3Li + {}^1_0n \rightarrow {}^3_1H + {}^4_2He$, $^3_1H + {}^{18}_8O \rightarrow {}^{18}_9F + {}^1_0n$.

22.3. 0.63 mm.

22.4. 3.0 sec.

22.5. 3.15×10^8 yr.

22.6. 23 20 yr.

22.7. 5.43×10^{-4}.

22.8. $N_{Rb}/N_{Sr} = (\epsilon^{\lambda t} - 1)^{-1}$.

22.11. 11.97 days.

23.1. 2.646 yr.

23.2. 16.3 μg, 5.8×10^{-3} cm.

23.3. Ir-192, Co-60, Cs-137.

23.4. 19,500 Ci.

23.5. 359 μg.

24.1. 227 times per year, $113 million; compare $70.6 million from fission.

24.2. 1.95, 0.51, 1.56.

25.2.   6.49 Ci.

25.3. 3.75×10^{19}/sec.

26.1.   1830 MW, 1350 MW; 26%.

26.2.   664 kg/sec, 2.6%.

26.3. 8.09×10^6 m², 366 J/m²-hr.

27.1. 1.41×10^4 mi², 1/8.

27.2. $2428, 55 yr.

Index

TITLES IN THE PERGAMON UNIFIED ENGINEERING SERIES